From reviews of the first edition:

"Patton's Oracle is a deeply personal account as the author recounts working with the humble, loyal and compassionate Koch as the old soldier is in a race against time, battling his terminal illness while attempting to complete and publish his book on tactical intelligence which he feels as his last professional responsibility."
—*Military Intelligence Professional Bulletin*

"He [Oscar Koch] had suffered the bad reputation of intelligence officers and often quoted General Omar Bradley on the subject: 'Misfits frequently found themselves assigned to intelligence duties.' Koch was the exception. Patton's Oracle explains why."
—*Journal of the American Intelligence Professional*

"When I finish a book with tears in my eyes, as I just did with a work by Robert Hays, Patton's Oracle: Gen. Oscar Koch, as I knew Him . . . I know I've experienced great writing. ... If you're a World War II history buff, Hays' book is must reading. It's a highly moving tribute to a great man by a friend whose life was enriched by the friendship."
—Huntington News Net

Other voices:

"I found that researching Patton through the eyes and actions of Oscar Koch was the most satisfactory approach to understanding Third Army in the Bulge."
—Peter Caddick–Adams (author, *Snow & Steel, the Battle of the Bulge, 1944–45)* to Robert Hays.

"Koch was the perfect G–2 for Patton, calm, deliberate, and just the right personality to interact with his volatile boss."
  —Carlo D'Este, (author, *Patton, A Genius for War*) to Robert Hays.

"Always he [Patton] had available in Koch's War Room the estimates of the situation of which was probably, in the field of intelligence, the most penetrating brain in the American Army."
  —*Patton, a Study in Command*, Hubert Essame.

Patton's Oracle

Gen. Oscar Koch, as I Knew Him

Biographical Memoir

by Robert Hays

Patton's Oracle: General Oscar Koch, as I Knew Him
A biographical memoir by Robert Hays 2nd Edition

Copyright 2013 by Robert Hays

No part of this book may be reproduced or transmitted in any form or by any means, electronic or mechanical, including photocopying, recording, or by any information storage and retrieval system without written permission from the publisher, except for the inclusion of brief quotations in a review or brief excerpts, properly credited, in research papers or other works of non–fiction. All rights reserved.

Excerpts from *G–2: Intelligence for Patton* are used by permission of Schiffer Publishing Company, which retains all rights.

Cover photo: Photo depicts Oscar Koch as Assistant Division Commander of the 25th Infantry Division in 1954. From MIPB: Military Intelligence Professional Bulletin, Michael P. Ley, editor. U.S. Army Intelligence Center (Fort Huachuca, AZ), publisher. Public domain, via Wikimedia Commons.

Library of Congress Control Number: 2023934255
1. Nonfiction. 2. Biography
ISBN 13: 978-1-950750-51-1
ISBN 10: 1-950750-51-5
2nd Edition printing, 2023
Thomas–Jacob Publishing, LLC Deltona, Florida
United States of America
Contact the publisher at TJPub@thomas–jacobpublishing.com

This book is dedicated to every man and woman
who has served in the Armed Forces
of the United States of America.

"Koch is the greatest G–2 in the U.S. Army. His record is without equal in every phase of intelligence."
—Robert S. Allen, *Lucky Forward*

Foreword

"Study history, study history. In history lie all the secrets of statecraft."

—Winston Spencer Churchill

General Oscar Koch is regarded as one of the most respected and impactful intelligence officers the U.S. Army has ever known, but for many years the true significance of his service to his country and his lasting impact on the broad field of intelligence was not widely understood outside of the Army intelligence community. This interesting and insightful biographical memoir honors and salutes this humble and proud American hero, and goes far toward closing that gap.

Oscar Koch's extraordinary military career began in 1915 and spanned service in the Mexican Expedition (General John J. Pershing's pursuit of the famed Mexican revolutionary and guerrilla leader, Pancho Villa) in 1917, World War I, World War II, the Korean War, and the Cold War. He retired in 1954 after achieving every rank from private to brigadier general during his nearly four decades as a soldier. He had served in units ranging from horse cavalry to the introduction of tanks and modern mechanized forces.

His most significant career role, as the reader will see, commenced when he was hand–picked as General George S. Patton Jr.'s G–2 (intelligence officer). In the summer of 1942, when he was a 45–year–old lieutenant colonel, Patton asked him to join his staff for what would be known as Operation Torch, the Allied invasion of North Africa. Patton knew Koch well. The two had become friends at Fort Riley, Kansas, and in 1940 Koch was named inspector general of Patton's 2nd Armored Division at Fort Benning, Georgia.

In the North African invasion, Koch served as chief of staff for sub–taskforce Blackstone, one of the three major units of Patton's 32,000–troop command. Not long after that he became Patton's G–2. For the rest of World War II, his meticulous attention to detail and implementation of processes became a force multiplier for Patton, and his superb and innovative performance in that critical intelligence role led to him being regarded as the best G–2 in the U.S. Army.

The contributions of Oscar Koch to the Army and the nation did not end with World War II. He remained on active duty through a variety of important roles, including founding and command of the Army's first intelligence school at Fort Riley, command of the 25th Infantry Division in Korea, and key Cold War intelligence assignments.

General Koch was posthumously inducted into the Military Intelligence Hall of Fame in 1993. Koch Barracks at Fort Huachuca, Arizona, home of the U.S. Army Intelligence School, was named in his honor the same year.

His Hall of Fame citation states: "BG Koch did more for

the development of combat intelligence than any other American intelligence officer, prior to or during WWII." Three paragraphs that follow list and enumerate the procedures and processes he developed and refined, many of which are still in use today.

In reference to General Koch's book, *G–2: Intelligence for Patton*, the citation declares it "may have been his greatest contribution to the Military Intelligence Corps. More than simply a memoir, the book was a tutorial for successful intelligence officers."

There can be little doubt that the movie, "Patton," released in 1970, had and still has a great deal of effect on the public perception of General George S. Patton Jr. It received great acclaim. But it has notable inaccuracies, and the author points out the great personal disservice it does to Oscar Koch.

He writes that Frank McCarthy, one of the producers, sent General Koch a letter enclosing portions of the movie script attributed to him as Patton's G–2 and asking him to sign off on the script. The inaccuracies were such that the general wrote back saying he could not approve these in their current form, but offered to make modifications. He got no response.

Lacking General Koch's essential approval, Patton's intelligence officer became a fictional character in the movie, without significant changes in the script. The opportunity to publicly proclaim and highlight the exceptional work of Patton's G–2 and his team was lost.

Robert Hays developed this fascinating biographical memoir during a series of personal meetings with General Koch in Carbondale, Illinois, based on the mutual respect of "soldier to soldier" conversations. He offers a deeply personal account of their relationship, and the character and nature of General Oscar Koch, especially his pride in and concerns about the intelligence field. He does a masterful job of revealing the character attributes which make the Oscar Koch story such a compelling one: grace, respect, discipline, humility, humor, and empathy.

Those character attributes inspired the confidence of General Patton and the entire staff, enabling Oscar Koch's impactful innovations as an intelligence officer. The reader will find this work to be a comprehensive account of those accomplishments and a fitting tribute to an exceptional American soldier and citizen.

—Maj. Brock Ayers, Military Intelligence Branch, U.S. Army (ret).

Chapter 1

THE GENERAL SMILED almost sheepishly as he handed me a letter that had just arrived a few hours earlier in the morning mail. His usual pleasant demeanor hardly masked a touch of honest curiosity. I took the envelope from him and saw it had been opened, a clean slit running the length of the spine. His name and address were neatly typed on the front and I was wondering why it merited my attention. Then I noticed the return address and hurried to see what was inside.

"What do you make of this?" the general asked.

I pulled out the contents of the envelope, a cover letter on high-quality business stationery and a few folded sheets of copy paper. I began to read. The letter was from Frank McCarthy, the Hollywood producer. McCarthy said he was making a movie about General George S. Patton Jr. and because the general had been an important member of the Patton staff, he was in it.

I was excited about the news. "Congratulations. You're going to be in the movies," I said.

"Maybe."

The general's approval was necessary, McCarthy informed him, and those portions of the movie script that included the lines for "Colonel Oscar Koch" were enclosed. Would he please look them over and send the movie-makers his authorization, in writing? And then this caveat: If he chose not to give the script his blessing, "we can fictionalize the character by substituting another name and making sure that the actor we choose to play the role does not look like you."[1]

The general looked somewhat uncomfortable. I knew he was a very modest man, but I knew, also, that he was immensely proud of his long association with Patton. He certainly would want to be in this movie. And by all means should be. What was the problem?

"Take a look at the script," he said.

I was surprised to see that lines of script attributed to "Colonel Oscar Koch" were few in number. In his post as G–2, the head of intelligence, Oscar Koch—long since promoted from colonel to brigadier general—had played a vital part in Patton's successes during World War II. He and Patton had been friends for many years, going back to a time well before the war began, and from then until Patton's death Oscar Koch had been a close and trusted confidant and advisor. He would be the first to label himself "a Patton man."

"That's not what I did," the general said, arching a finger toward the script. "There's no way I can approve this."

As I looked over the skimpy segments of movie script McCarthy had offered, I saw at once that the lines grossly distorted Koch's real-life role with Patton. Further, without the rest of the script to put the Koch dialog into context, it was impossible to envision precisely how he would be characterized. But I'd seen enough to make me angry.

"This line makes you look like a fool," I said.

He laughed. "Maybe they know me better than we think."

"Like hell. Patton didn't countenance fools, not from what you've told me. What are you going to do?"

"I'll write this man and tell him I can't sign off on this but I'm willing to help fix their script if he's interested. Is it all right with you if I offer to send along parts of our book, if he'd like to see it?"

I said, "Of course. Give him anything he wants. You ought to be in this movie."

A few days later, the general wrote a polite response to Frank McCarthy. He told the producer he could not approve the script as it was and offered to help correct it. He never heard anything more. True to McCarthy's word, Oscar Koch was replaced in the movie by a fictitious character named "Colonel Gaston Bell."

I'm happy General Koch confided in me that day and shared the content of Frank McCarthy's letter. Otherwise, I might never have known why he was not portrayed in the 1970 movie, "Patton," which was very good in spite of some glaring factual errors and other omissions. The general and I were friends, and at the time we were collaborating on a book on combat intelligence that eventually was published under the title, *G–2: Intelligence for Patton*.[2]

We were an improbable team. I was only nine years old in December 1944, when Oscar Koch earned authentic World War II hero status as he stood virtually alone in warning Allied military leaders of the coming breakout by German Army troops in what would come to be known as the Battle of the Bulge. There is ample evidence that had

General Dwight Eisenhower and General Omar Bradley acted on Koch's information, the lives of thousands of American soldiers might have been saved and the war in Europe might have been shortened significantly. Patton was the only one to take his warning seriously and plan accordingly.

The U.S. Third Army was ready to move when Eisenhower gave Patton the green light.

Oscar Koch had chosen to be a professional soldier and spent almost four decades in the U.S. Army; I had been drafted at age twenty and couldn't wait to serve my time and return to civilian life. He retired from military life a year before I first put on the Army uniform. It was mere chance that led us to cross paths. In spite of our differences in age and background, though, we developed a deep and abiding friendship. I came to respect, admire—and yes, love—this man, who influenced my life in more ways than I can count.

I was granted only four years to share life with the general, a period that was far too short. In the beginning he lifted my spirits as we joined in a common purpose. In the end, I endured the anguish of watching an insidious cancer purloin the life from his body even though he never would surrender his gallant spirit. But what a remarkable four years it was, how grateful I am to have had that privilege!

I met General Oscar Koch early in 1966, a trying time in America. Opposition to the war in Vietnam, where more than two hundred thousand U.S. troops were involved, appeared to grow more intense almost daily and open racial strife had become virtually commonplace in much of the nation. Network television news reports brought both the war and the racial disturbances into the homes of viewers.

Intense opposition to the draft led to open defiance of a new federal law against draft-card burning.

Conservative opinion writers railed against the protests. On the other side were news media such as the *Capital Times* in Madison, Wisconsin, which carried an April editorial calling for editors and law enforcement officials to stop trying to create "an atmosphere of war hysteria in which emotionalism and chauvinism are substituted for reason and patriotism." In defense of the antiwar protestors, the newspaper said millions of patriotic Americans considered the nation's Vietnam policy a tragic mistake. "In this country," it reminded readers, "they have a right to speak their views and demonstrate for them—as much right as those who think our Viet Nam policy is sound."[3]

American citizens were nearly evenly divided in terms of support for and opposition to the war. Yet there still was immense sentimental backing for the soldiers. Sergeant Barry Sadler's recording of the touching "Ballad of the Green Berets" became the most popular song in the country in early March and held that position for more than three months.

The general and I made a conscious effort not to be sidetracked by the social turmoil going on around us. This was difficult, and we did not ignore it. We simply were careful, as he expressed it one day, "not to waste too much time discussing how we could fix all the problems in the world if the powers that be only would listen to us." It was important that we kept our eyes on our own goal, the completion of a book on intelligence in combat.

I was a special correspondent for the *St. Louis Post-Dispatch* (which means I wrote for the newspaper on a free-lance basis) and early in our relationship I had persuaded the general to let me write a feature about him and his wartime experiences under Patton. It hadn't been an easy sell. I think my argument that his story would only enhance Patton's reputation was the one that eventually won him over.

The article turned out well, but when I read it today it strikes me as painfully incomplete. I know a great many things about General Koch now that I wish I had known then. High on this list would be his masterful use of ULTRA, the secret British intercepts of enemy radio messages after they'd broken the German code. But at that time even the existence of ULTRA had not yet been revealed, and Oscar Koch was not one to go against restrictions on what he should talk about. I also wish I had known he spent two years near the end of his military career on a virtually clandestine assignment to the State Department helping to strengthen the work of the CIA and took special assignments after he retired. This one may not have mattered, though, as he downplayed the significance of that critical tour of duty even after I learned about it some months later. I doubt he would have been particularly forthcoming on that experience, either.

This is not to say he was reluctant to answer my questions. It merely is an acknowledgment of the fact he didn't particularly like to talk about himself, especially when someone wanted to make him sound "important." He was a team player, and in his mind no individual should outshine the team. Except for the captain—Patton himself.

My first meeting with the general was little more than happenstance. I was a new and terribly green public relations writer for Southern Illinois University in Carbondale, where the chancellor, always sensitive to town–and–gown relationships, had promised to help a local civic group promote its activities. This meant the services of someone from the PR office, and to my very good fortune that someone turned out to be me. Merely the luck of the draw—or maybe I was the only one not busy that day.

In any case, I was to meet a retired brigadier general at the home of a war widow, Mrs. Burrell Smith. Mrs. Smith wanted to present the flag that had draped her husband's coffin to a national flag bank organized by the Carbondale Memorial Day Association. Her husband was a veteran of World War I.

The retired one–star general turned out to be Oscar Koch, the association's executive director. His work with the association was voluntary and, I would learn in the coming weeks, typical of the way he chose to spend his time. He had moved to Carbondale in retirement because it was the family home of his wife, Nannie Caldwell Koch.

The general and I arrived at exactly the same moment at Mrs. Smith's home, a modest, well–kept white clapboard house which sat squarely between one of Carbondale's old, brick–paved streets and the embanked right–of–way of the Illinois Central Railroad tracks. He introduced himself and joked that our timing was perfect; we might be "kicking off a military operation." He admired punctuality. We were off to a good start.

Mrs. Smith was somewhat misty–eyed as she presented General Koch with her flag. But she said she would be grateful if it could be put to some proper use "rather than just be hidden away in a dresser drawer until I die." He clearly was impressed by the reverence with which she handled the flag and promised that, beginning on Memorial Day, it would fly over the city's Woodlawn Cemetery. He told her the flag eventually would be disposed of in a proper manner.

The general asked about her late husband, James Benjamin "Candy" Smith, a major league baseball player with the Chicago White Sox and the Detroit Tigers. Her husband apparently had been a great storyteller, and she repeated for us some of his best anecdotes. Most of her stories were from his days with the Tigers when that team was managed by the famed Ty Cobb.

The general laughed heartily and I could see this helped put Mrs. Smith at ease. What had begun as a solemn occasion turned into a pleasant visit that stretched into more than a full hour. I took photos of the flag presentation and wrote a news story to be distributed by the university news service. Both the news report and a picture were picked up by several area newspapers. The general let me know he was pleased with the publicity for the flag bank and, more important, for the Memorial Day Association.

Before this assignment, I never had heard the name Oscar Koch. I knew nothing about him beyond the fact he was a retired U.S. Army general. This was daunting enough. As a draftee, specialist third–class regimental special orders clerk, my military service had not included much rubbing of elbows with generals. I seldom had contact with anyone who ranked above captain, and had I not been responsible for maintaining the duty roster for field–grade officers—majors through colonels—I hardly would have been aware we had all this higher brass in the regiment.

The one occasion on which I did associate with the battalion commanders turned out a bit awkwardly. I was on charge–of–quarters duty and had to call each of them individually at 3 a.m. and advise them of a post–wide alert exercise. One colonel asked, "What the hell am I supposed to do?" I didn't know. For another one I had a wrong number and woke a stranger in the night. She was not pleased.

None of this did much to salve my pervasive unease with high–ranking officers.

If I had known that Oscar Koch had been a key member of Patton's staff, I might have been even more intimidated. Nobody thought of Patton as a pussycat. Why should I expect one of his hand–picked staff officers to be any different?

I couldn't have been more wrong.

General Koch was a robust man, immaculately groomed, and he wore a broad smile. I was surprised by his pleasant, down–to–earth manner. This was a general? I didn't know it then, but Oscar Koch was a soldier who rose from private to brigadier general the hard way, passing through every intervening enlisted and commissioned rank that existed at the time. He clearly had earned his star.

He soon learned I was a veteran, too, and readily accepted the fact that we had this in common. He honored those who had served; rank apparently was not especially important.

I had met a great many interesting people in my work as a newspaper reporter and university public relations writer. None had been more interesting than Oscar Koch. I found him a joy to work with.

This was not a man who lived in the past. He faced his days with good humor and exuberance and, although I think he would have been content wherever his feet touched the ground, he relished the tranquility of small-town life and the serenity he found in the natural beauty of the rolling southern Illinois countryside with its miles of peach and apple orchards and nearby Shawnee National Forest.

General Koch was my senior by many years, but nonetheless as energetic as I was and ready to devote himself to virtually any worthy cause. He was, as Patton might have said, "a doer, not a talker."

Our relationship was positive from that first meeting. There were more projects coming up that the Memorial Day Association needed publicity for, and the general apparently reported back to the chancellor that he would be pleased to continue to work with me. For my part, I was grateful I had been volunteered.

I spent a great deal of time with Oscar Koch in the coming weeks and did what I could to help his organization in its drive to gain recognition for Carbondale as the site of the nation's first formal Memorial Day celebration. This claim was based on an observance that occurred in 1866 in the city's Woodlawn Cemetery, which holds the graves of twenty-two Civil War veterans. Monuments in the cemetery, a peaceful, wooded hilltop, identify both Union and Confederate troops and include one that marks the final resting place of an unknown soldier.

The speaker that day a hundred years earlier was General John A. Logan. General Logan was a native of Jackson County, where Carbondale is located, and a national Civil War hero. He gained a reputation as one of the Union Army's most capable leaders and after the war went on to become head of the Grand Army of the Republic, an influential national veterans' organization.[4] Two years after the Carbondale ceremony, General Logan issued his famous GAR General Orders No. 11 calling for May 30, 1868, to be a day devoted to "strewing with flowers or otherwise decorating the graves of comrades who died in defense of their country during the late rebellion, and whose bodies now lie in almost every city, village, and hamlet churchyard in the land."

His proclamation was broadly accepted and led to widespread observance of what at first was called Decoration Day. This later was changed to Memorial Day.

As a highlight of its promotion, the Carbondale Memorial Day Association organized a centennial observance of the 1866 Woodlawn Cemetery event. Most of the planning load was carried by General Oscar Koch. The work hardly could have been in more capable or more willing hands.

On a beautiful spring morning, color guards led a parade through town to the cemetery, where there was a wreath–laying at the grave of the unknown soldier and a formal reading of Logan's 1868 GAR General Orders No. 11. The featured speaker was Chaplain Ernest C. Klein of the U.S. Army Reserve 12th Special Forces "Green Beret" unit, a combat infantryman and senior parachutist with the 95th Division in World War II. Clyde Choate, a member of the Illinois General Assembly and World War II Medal of Honor winner, presented a flag to the national flag bank.[5]

I would have been there anyway, of course, but I was hired to report on the day's activities as a correspondent for the United Press International wire service. I took such stringer assignments from both the UPI and the Associated

Press because I hated to turn down the money, and also because I wanted to stay on their good side. The wire services were important outlets for our PR efforts and we tried hard not to alienate them. The money wasn't much; on this day I was paid six dollars for a half-day's work. The AP paid about the same.

The celebration featured performances by a church choral group, a high school band, and the Southern Illinois University Concert Band. The music was at the same time somber and stirring, appropriate to the occasion, a tribute to those who had given their lives for our country and a call to patriotism. I thought about the cheerful tone of a recording by Herb Alpert and the Tijuana Brass I had been listening to earlier on the car radio and marveled at the wondrous suppleness of music, which surely has something fitting to offer in any circumstance.

A bugler and a ceremonial firing squad brought a stunning end to the centennial reenactment. I saw tears in the eyes of many in the modest-sized crowd of people, mostly the older citizens, who had come out to witness the commemoration. And pride on their faces. These were people who loved their country and honored the men and women of the armed forces who had been called to its service. They would have known about Patton, but I wondered how many were familiar with the remarkable record of Oscar Koch.

The Woodlawn Cemetery centennial celebration struck me as beautifully serene and in every sense fitting. I sought out General Koch as the ceremonies drew to a close, expecting to find him exhausted after his busy day. If he was tired at all, it didn't show.

"Was it good?" he asked.

"Yes, Sir. It was good." And I truly believed it was.

This story might end here were it not for John W. Allen, who was president of the Memorial Day Association. He was close to the general in age and the two of them had been good friends for some years, beginning well before they got involved with the Memorial Day Association.

I knew John Allen from my days as an undergraduate journalism student. A past Illinois State Historical Society president and Illinois state "Historian of the Year" in 1955, he was a fascinating storyteller whose weekly columns on regional history were distributed by the Southern Illinois University news bureau. I worked there part time, and one of the pleasures of my job had been editing his colorful stories. He never failed to express his gratitude for my modest contribution and showed me the same level of respect then as he did now that I was a professional staff member in this same unit.

I think it would have been some time before General Oscar Koch, reserved as he was when talking about himself, would have told me the things John Allen had learned through their long relationship—things which John Allen, in turn, told me. These included the fact that the general had saved his own intelligence reports from the Patton commands and had them stashed away in old Army footlockers in his basement. On retirement from the Army he'd been granted a Guggenheim Fellowship to support research for and writing of a text book on combat intelligence. He studied documents in the German military archives to see how the information in his reports measured up when compared to the actual German records. The general had written a book on combat intelligence, but the manuscript had been rejected by several publishers. He'd recently been back at work on revisions and still hoped to see his book in print.

None of this surprised me. I had come to recognize that Oscar Koch was a studious man who believed strongly in the importance of the field in which he specialized. What I did not realize as yet was that it reasonably could be argued he was the best intelligence officer in U.S. Army history.

John Allen's revelations had piqued my interest in Oscar Koch even more. I was eager to get started on the article I had promised the *St. Louis Post–Dispatch*. This modest, unassuming man plainly was an unsung hero of World War II and it was high time he gained some recognition.

Chapter 2

ALTHOUGH IT WAS years ago when I last visited there and sat at that crowded dining room table, trying to sort out the big picture and meticulously poring over even the smallest details with the general, my memories of the Koch home still are quite lucid today. I can see it in my mind's eye, an appealing country place upon which the city was rapidly encroaching. I had driven by it many times before I knew who lived there and it had become one of my favorite Carbondale residences.

The house was modest in size but graceful in style. It sat close to a highway, backed by well-manicured grounds that slanted away toward a floodplain and tree-lined creek. A gravel driveway ran alongside—Nan Koch, in her words, "hated concrete"—and led to a carriage house that sheltered a gleaming black Ford sedan. An ancient silver maple tree shaded the carriage house and much of the home. There was a tall steel flagpole in the front yard, where the American flag flew from early morning until sunset every day during clement weather, prominently visible to passersby on the highway.

When I went inside the house the first time, I was

struck by its simple elegance. This was a home which said a great deal about the remarkable couple who lived there.

Reminders of Patton were everywhere. The living room furnishings were arranged so visitors would focus on an open–hearth fireplace. A highly polished brass desk clock that at one time had belonged to Patton adorned the mantle, and on the wall above it was fixed a gleaming Patton Saber, the straight, double–edged cavalryman's sword designed by the legendary general, adopted by the U.S. Army in 1913, and carried by countless mounted troops over the years.

General Koch noticed me gazing at a life–like oil painting of Patton that dominated one wall. "It's a pretty good likeness of the Old Man," he said affectionately.

I nodded in agreement. I would have to take his word for this, because I had no clear notion of what George S. Patton Jr. looked like. The man in the painting was rather handsome and wore the uniform sharply.

The house was immaculately kept. Nan Koch, I soon learned, would not allow disarray. The lone exception was the dining room table, which always seemed to be piled high with the general's working papers. But she was in command here, and I think that's exactly the way the general wanted it. It was as if he already had carried enough responsibility to last him for a lifetime.

The road in front of the house is Old Illinois Route 13, an outdated two–lane concrete highway which even then already had been replaced with a new version that tempered the steep hills and sharp curves of the original. Only a few hundred yards to the east, the old joins the new and becomes West Main Street of Carbondale, Nan Koch's home town. She still thought of herself as the former Nannie Caldwell and the Caldwells were a pioneering Carbondale family. The Koch house and surrounding grounds were part of the original Caldwell estate.

Nan told me that Orchard Drive, which angles across one of Carbondale's better residential areas, was named for her family's apple orchard. (I'll always remember it for a different reason: Our first-born son, Alan, a precocious four-year-old who had learned to read by watching an adult literacy series on a Paducah, Kentucky, television station, insisted the "Orchard Dr." on the street signs meant "Orchard Doctor.")

Carbondale lies deep in southern Illinois, slightly offset from the center of a giant wedge of land bounded by the Mississippi River on the west and the Ohio River on the east and transected from river to river by a spine of the Illinois Ozarks marked by outcroppings of sandstone and limestone cliffs. It is parallel in latitude to Richmond, Virginia, closer in driving distance to Huntsville, Alabama, and Oxford, Mississippi, than to Chicago, and about equal in distance from Chicago, Little Rock, and Chattanooga.

As reflected by the monuments on the graves of the Civil War dead in Woodlawn Cemetery, the roots of many of the area's early settlers lay in the South.

This included the Caldwells. Nan Koch proudly proclaimed that Caldwell County, Kentucky, was named for one of her ancestors. This would have been John Caldwell, the second lieutenant governor of Kentucky and a veteran of the 1786 George Rogers Clark campaign against British forts on the Ohio and Mississippi rivers. Clark and other prominent men of his time thought the British were aiding the Indians in their attacks on settlers in Virginia and Kentucky. Clark's campaign centered on the region that would become southern Illinois, near the site of present–day Carbondale.[1]

One bit of family history that no doubt struck Nan as ironic in later years involved her brother, Isaac. He received an appointment to West Point but turned it down. Had he accepted and finished on schedule he would have been a classmate of Dwight Eisenhower and Omar Bradley.[2]

But well beyond her prominent family name, being married to Oscar Koch had given Nannie Caldwell higher standing on a wider world stage. She was a general's wife. Oscar Koch had retired from the U.S. Army as a brigadier general after nearly forty years of service, culminating in a brief period as commander of the highly decorated 25th Infantry Division, the celebrated "Tropic Lightning," near the end of its thirty–seven month tour during the Korean War.

It seemed to me that, his distinguished military record notwithstanding, casual visitors to the Koch home would not have presumed the man who lived there to be a soldier had it not been for the extensive collection of military memorabilia. Oscar Koch was a spirited, somewhat dapper man who was polite and considerate, more scholarly than many of the professors at the big university that dominated his community. He might have passed for a lawyer or doctor of medicine. Or a retired businessman whose roots, like Nan's, ran deep in the community.

As I sat in the comfortable Koch living room and began an interview for the *Post–Dispatch* article, I was struck by the general's personal story as well as by the details of his military experience. Nan sat in on the session and readily shared memories of their life together. She was proud to have been an Army wife. I asked what year they married.

"It was 1924—wasn't it, Oscar?"

"I know it was a long time ago," the general said. And to me, "Don't ask her if she would do it again."

Nan smiled sweetly and he patted her hand. I did not ask if she would do it again, although the answer seemed clear.

"Oscar got a twelve–dollar and fifty cents monthly maintenance allowance for his horse," Nan said, "and eighteen dollars for me. As far as Uncle Sam was concerned in those days, horses were just about as important as wives." Then she laughed and added, "I think in some cases, but certainly not in his, their horses might have been more important."

The general laughed, too, though I had a feeling he'd heard this tease before.

At that time, General Koch explained, there were some twelve thousand officers in the U.S. Army, eight hundred of them cavalry. The Army encouraged cavalry officers to buy and maintain their own mounts, with the allowance Nan had just mentioned. He had been what he called "an old–fashioned horse cavalryman" since the day he joined the Light Horse Squadron Association in his home town of Milwaukee in 1915, five months after his eighteenth birthday.

I tried to recall what I might have been doing at 18. There was nothing worth remembering.

The general was reluctant to tout his personal accomplishments, but it was obvious he relished the occasion to talk about life with Nan. Theirs had been a typical military life and that meant a life on the move. Each would interrupt the other as they thought of favorite stories. One transfer they both remembered well took place in 1939, when he was reassigned from Fort Riley, Kansas, to Fort McDowell in San Francisco. The Army routed them and their household and personal belongings first to New York City, then by ship to California by way of the Panama Canal.

I'm sure they told me how long the journey took, but I don't remember. And looking back at my notes now, I find no mention of their time en route. (Yes, I'm a packrat; I save things like interview notes forever, stuck away in banker's boxes in our crowded garage, and fortunately so. Without those old notes this book would not have been possible.)

My notes highlight the fact that Oscar Koch left Kansas with an important new friend. It was during duty at Fort Riley that he met George Patton. When Patton was named commander of the newly organized 2nd Armored Division in August 1940 he enlisted Koch to join his staff at Fort Benning, Georgia, as the division's inspector general. Oscar Koch would serve under Patton for a large part of his remaining military career.

Rather than talk about himself, the general carefully laid out for me the state of military intelligence in this pre—war period.

Succinctly, intelligence was not given the priority it should have had. I quickly grasped the depth of his feelings on the matter, and it was clear he spoke from a true appreciation for the importance of good intelligence to a commander in combat and not from personal bias.

The general explained that American forces in World

War I had relied heavily on their allies for intelligence information, and during the between–wars era there was no great outcry for intelligence activity. Many officers, assuming that intelligence staff duty would do little to advance their careers, were much more eager to get command assignments on their records. He pointed to General Omar Bradley, who as 12th Army Group commander was Patton's superior, as an example. In his book, *A Soldier's Story*, Bradley writes that, "Misfits frequently found themselves assigned to intelligence duties. And in some stations G–2 became the dumping ground for officers ill–suited for command. I recall how scrupulously I avoided the branding that came with an intelligence assignment in my own career."[3]

This attitude was extremely unfortunate, in General Koch's opinion, although he recognized that it was common at the time. Given its low priority, not nearly enough attention had been paid to intelligence training in the years leading up to World War II.

"We paid a price, too," he said. "Intelligence officers are made, not born."

I would learn over time that this was an Oscar Koch mantra. It guided him in his fervent desire to see the American armed services launch and maintain effective training programs in combat intelligence, a dream that was to become a personal responsibility after the war when he was appointed the first head of the new intelligence department at the Army's Ground General School at Fort Riley.

Unlike most of the top commanders in the war, Patton had been an intelligence officer. He served three years as G–2 for the Hawaiian Division, where his activities were so poorly funded he tried supporting the intelligence section with his own money. His dedication wasn't appreciated, apparently, and he was told he was a fool for doing so.[4]

Patton warned in a report dated June 3, 1937, that Japan was willing and quite possibly able to launch a surprise attack on Pearl Harbor. The reasoning he spelled out in that document would prove to be right on target over the next few years.[5] He likely had little doubt war was coming when he took command of the 2nd Armored Division three years later and his determination to train his troops accordingly may have led, at least indirectly, to his famous nickname, "Old Blood and Guts."

As the story was told by General Koch, at one of the weekly conferences of division officers Patton closed his discussion of administrative matters with, "And now I will get to my favorite subject: blood and guts." He was going to talk about a renewed emphasis on the division's combat readiness and the tough training the men could expect. Patton had been in combat and saw no reason to sugarcoat it.

"After that," Koch said, "it wasn't unusual on conference days to hear someone say, 'Let's go on over and see what Old Blood and Guts is going to talk about today.' The nickname just seemed to stick. Sometimes Patton pretended not to like it, but I think he secretly got a kick out of it. But all in all I think he liked it best when his soldiers referred to him simply as 'the Old Man.'" Regardless of his own preferences, though, "Georgie" was the name most commonly used by Patton family members.

Patton's battle tactics during 2nd Armored Division maneuvers attracted attention. In July 1942 he was selected to command the Western Task Force, which was to join the British for an invasion of North Africa in what would come to be known as Operation Torch. He'd been preparing troops with exercises at the Desert Training Center, a vast military facility that sprawled across parts of California, Nevada, and Arizona, and then returned to the East Coast.

At an early opportunity he stopped by to visit his former command. He invited several of his old friends to dinner

Oscar Koch, then a lieutenant colonel, had reached the age of 45—too old for combat under existing military regulations except as a volunteer. But if Patton was going to war, he wanted Koch to be there with him.

As General Koch easily recalled, "After dinner Patton called me out on the porch and said, 'Koch, do you want to go to war?' I said, 'Yes, Sir.' He said, 'Well, you're going.' And that was that."

"Did you actually want to go?" I asked.

"Yes. I was a soldier and there was a war that had to be fought. And I couldn't think of a better way to do it than to fight under General Patton. I knew he was the best."

The Allies launched the invasion of North Africa the following November. The invasion plans were relatively simple. Patton's American forces, numbering some thirty-two thousand troops, would land on the Atlantic coast, with Eastern and Central task forces mounted in the United Kingdom set to invade from the Mediterranean.

Three sub–task forces under Patton would include Blackstone, commanded by General Ernest N. Harmon, with Lieutenant Colonel Koch as chief of staff.

As I came to know Oscar Koch, to see how thorough he was, how careful with every detail, I could well imagine he was an extremely capable chief of staff to General Harmon.

Attention to detail also is a quality needed by an intelligence officer. Four months after American troops landed on the beaches of French Morocco, Patton called on Koch to be his G–2, the head of his intelligence section.

I tried to do justice to the general's distinguished military career in the *Post–Dispatch* article, published as a full-page feature in the newspaper's "Everyday Magazine" section under the banner headline, "He Helped Decide to Hold Bastogne."[6] With all the ground there was to cover, I had emphasized Oscar Koch's role during the Battle of the Bulge.

The German Army's all-out offensive in December 1944 quickly put the small Belgian town of Bastogne, a crucial transportation juncture, square in the path of the enemy advance. On the night of December 19, or perhaps early on the morning of December 20, Patton and General Omar Bradley faced a decision as to whether the Allies should fight to hold Bastogne against the onslaught of enemy forces. Patton, as he commonly did, conferred with his intelligence chief. Colonel Koch said yes. Although there would be a concern about supplying the troops, Bastogne's location made it critical to the effort to repel the enemy's rapid thrust.

Bastogne, of course, came to be one of the most recognized appellations of World War II. When Patton's forces broke through the day after Christmas to rescue troops of the 101st Airborne Division and other units surrounded there by the Germans, his stature as a hero was greatly enhanced among the American people. Had they known Patton's success was made possible by the brilliant performance of his intelligence chief, Oscar Koch might have gained recognition as well. In any case, holding Bastogne proved to be a major factor in breaking the back of the German attack.

General Koch said he never knew precisely to what extent his recommendation influenced the decision to hold Bastogne, a fact I reported in the article. Bradley was Patton's superior, and there's little evidence that he appreciated the role of intelligence to the extent Patton did. From what I know now about Koch's relationship with his famous commander, however, I'm confident that had the decision been Patton's alone Koch's input would have settled the question.

Even though he had been a somewhat reluctant subject, General Koch was pleased with the *Post–Dispatch* article. He heard from old friends long removed who'd read it and said they were glad to see him gain some well–deserved recognition. He still wasn't especially happy to be in the limelight himself, but I'd done my best to give him due credit for his brilliant intelligence work for Patton and I think he was grateful.

Chapter 3

Not many people in and around Carbondale knew the full story of Oscar Koch's background. Some were aware he'd served under Patton; most were not. His name was more likely to be recognized for other reasons, strictly local in origin. He did those things expected of leading citizens—trustee of the First Christian Church, director of the YMCA, member of the Rotary Club and the Elks.

He was active in the Jackson County Historical Society and had been appointed to the Carbondale citizens' advisory board.

Those who did know him understood that the general's contributions always went well beyond mere lip-service. His work was impressive. No detail was too small to get his attention, none large enough to intimidate him. He liked to be called on when there were things to be done, because it simply wasn't in his nature to sit around and do nothing.

His civic affiliations were not important to me, but his reputation as a man of his word was. He was polite and considerate and he soon became a favorite visitor at our house.

When the general telephoned one morning to check on

my schedule, my wife Mary answered the phone and he asked to speak to "your daddy." She still had the pure South Carolina midlands accent of her native Columbia, spoken in a soft voice, and it wasn't unusual for callers to mistake her for one of our two young sons. He knew what he'd done before I came on the line and asked me to apologize. She wasn't offended, of course, but he was embarrassed.

General Koch never liked to make the same mistake twice. But he did. A week later the same thing happened a second time.

"I did it again!" were his first words to me. There was such exasperation in his voice that I felt sorry for him. I could have guessed the cause of his dismay, but he quickly went on to explain: "I asked Mary again if I could speak to her daddy. She must think I'm senile."

I assured him she thought nothing of the kind. I could have told him she admired him immensely, not only because of his record as a true American hero but also because of his gentle, grandfatherly presence. Mary didn't always take readily to newcomers, but with the general she had been somewhat smitten from the outset. I always found her almost infallible as a judge of character.

When I went to the Koch house, the general never failed to ask about Mary and the boys, and I never doubted his concern was genuine. His friends were like family to him and he treated us as such.

There's plenty of evidence that the Oscar Koch we came to know was the same Oscar Koch seen by those who knew and worked with him during his decades of military service. Contemporary descriptions paint a picture of a very considerate man, regardless of the circumstances. Stanley

P. Hirshson, in his marvelous biography, *General Patton: A Soldier's Life*, quotes Lieutenant Colonel Bernard S. Carter, a Koch assistant, in a letter to Carter's wife: "He [Koch] really is a wonderful person—he's just good—very thoughtful & kind & unselfish & modest ... all his subordinates are devoted to him."[1]

Colonel Ralph M. Luman, who served with Oscar Koch at various times, wrote that from the time Koch joined the 2nd Armored Division through the planning of the African and Sicilian invasions, "he was always an inspiration to the officers and men of his G–2 Section. We admired his professional abilities, and loved him for his personal warmth and charm."[2]

The relative to whom Koch was closest and no doubt the one who knew him best was a cousin, Dr. Murray Zimmerman. Dr. Zimmerman was a Los Angeles dermatologist and a forty–year member of the faculty at the University of Southern California School of Medicine with a charismatic firebrand, "take no prisoners" personality. He supposed that for a thirty–year–old like me the general "must have been a formidable guy to work with." But I needn't have been concerned, he added, because Oscar Koch was "kindness personified." He described the general as "a self–effacing, quiet, modest, diffident guy."[3]

This was, indeed, the Oscar Koch we knew.

From the very beginning, the time I spent with the general was pleasant time. He had a subtle but pervading sense of humor and could be clever with words.

Because I did not know the German language, he had fun "teaching" me things I needed to know, beginning with a dire warning. If I traveled to Germany I should be careful not to make the blunder Americans often did, mistaking *damen* for "men" and *herren* for "women." This could lead to embarrassing consequences.

I told him if I ever planned a trip to Germany I'd draft him as my guide. He said he'd be happy to volunteer.

While this is the side of Oscar Koch I'll always remember best, I will not lose sight of the fact that this quiet, benevolent man was by choice a professional soldier. He had seen America's wars from the front lines and, while it was apparent he took no pleasure in the horrors of battle, it also was fully evident he held the profession of arms in high regard and understood there are battles that must be fought. And if fought, won.

Koch's first experience in live warfare came with service on the Mexican border under General John J. Pershing during the American "punitive expedition" of 1916–17 chasing the elusive Mexican revolutionary leader, Pancho Villa. He spent 14 months in France during World War I and saw three years of continuous service in World War II, facing combat in French Morocco, Tunisia, Sicily, and across the body of Europe. He confronted momentous decisions that meant life or death for thousands of troops in battle and a large part of his service was with Patton, by any measure one of the most demanding and hard–driving military leaders in history.

I don't know whether General Patton and others with whom Koch served were aware of it, but one small detail on his record was, well, a bit of a lie.

While his Army file carried the name "Oscar W. Koch," he actually had no middle name. The initial was a creative expediency resulting from a young man's quick thinking at the time of his first run–in with higher military authority.

"I had a grandfather named Oscar Koch and a father

named Oscar Koch Jr.," he told me, "so I thought I should be Oscar Koch the Third." This was the name he put on his application for enlistment in the Light Horse Squadron Association in Milwaukee. The application called for last name first and, always careful to follow instructions, he wrote "Koch, Oscar III."

A gruff line sergeant sitting in the office when he showed up with his application demanded to know what the initial stood for. Young Oscar was afraid to tell him it was not an initial, but a Roman numeral.

He still appreciated his own cleverness a half-century later, although he admitted the joke was on himself. "The good desk sergeant thought it was a W, and I didn't want him to think I was trying to get smart with him so I said 'William' real quick," he explained. "I've been Oscar W. Koch ever since."

Middle initial or not, pronunciation of his name often was an issue. Nan told me that during his months in combat he had a sign above the door of his field tent that read, "I have bourbon, I have Scotch. If you drink with me, call me 'Kotch.'" Without the benefit of this unadorned but effective guidance, some called him "Cook," after the name's German roots. Some said "Coke."

Oscar Koch—without the W—was born in Milwaukee on January 10, 1897, into a Jewish German–American family. Some military historians have made a great deal of the fact he was a Jew, primarily because of a perception that Patton was anti–Semitic. This has presented them with something of a dilemma, because Koch was one of Patton's closest friends and confidantes as well as a trusted advisor.

The general and I never discussed all this. He did not

practice the Jewish religion and, in my experience, never appeared to take his Jewish heritage seriously. In fact, it never occurred to me he was a Jew. I don't pay much attention to such things and would not have cared one way or the other, but surely if his "Jewishness" had been as obvious as some have made it out to be, I might have noticed. I suspect that Patton looked on him much the same way I did.

Koch's German background was important, however, because he was fluent in the German language and this clearly was an advantage. He translated General Maxmilian Von Poseck's book, *The German Cavalry in Belgium and France, 1914*, a detailed account of operations in the opening campaign of World War I, for the U.S. Army War College. After that he was called on to translate other German military documents higher authorities thought might be put to good use if the expected war with that nation actually came to pass.

It was easy to get caught up in lively conversation with the general and be slow to get around to the questions I wanted and needed to ask. Given that he was not one to initiate much discussion about his military career, I didn't know about many of his stops along that route until I managed to get an abbreviated copy of his resume. His record reads like an adventure story that covers the greater part of four decades. And to say the least, the road he travelled often was a bumpy one.

He faced a challenging obstacle at the very beginning—one that might have ended his service in the armed forces before it began. It was not an enemy challenge, but a friendly one. I asked him to tell me about it. This appeared to be one personal revelation in which he took some pleasure.

When he set out to enlist in the Milwaukee cavalry squadron, he found he wasn't old enough to join without parental consent. His widowed mother was reluctant to give permission. Fighting was raging in Europe and many believed it was only a matter of time until American forces would be involved. Like any mother, she had no wish to see her son go off to war.

The young Oscar knew he would need help to persuade her, and called on a Chicago cousin, William Zimmerman, who was a few years older. His cousin apparently had a good deal of influence with his mother and after the two of them joined forces they eventually wore her down. But they still needed something to clinch their case.

"I think I finally won her over with the simple argument that I probably wouldn't see any action, anyway," he told me, evidently recognizing I would appreciate the irony.

"You sure misjudged that one," I said. "How did she feel about it later?"

"We never talked about it again."

"No guilt?"

I remember his facial expression as he answered, "Boys always have guilt when they mislead their mothers." He still felt as if he had betrayed his mother's trust.

In any case, the Light Horse Squadron Association offered a taste of military regimentation from the beginning. It was an active unit that drilled once a week. A private was paid $15 a month. Looking back, the general allowed that this was a respectable figure for the time, even though the squadron members had to pay dues. The money they paid back went toward maintenance of the armory and equipment, setting the stage for his later experience as a

U.S. Army cavalry officer.

And probably most important, he added, "I made contacts that were to become lifetime friendships." He still could tell me the names of many of those young cavalrymen who rode beside him.

Oscar Koch—now, Oscar W. Koch—hadn't been there long before the horsemen's unit got a prestigious assignment. President Woodrow Wilson visited Milwaukee in early 1916 on a highly publicized national speaking tour to promote American military preparedness. The squadron was called on to form a mounted honor guard to accompany the president from the Hotel Pfister to Milwaukee Auditorium and back. This surely was a heady experience for one of the unit's newest cadets.

But ceremonial duties and weekly drills soon were replaced by the real thing. The action he'd assured his mother he was not likely to see began a year after he joined his home-town unit when it was re-designated Troop A, 1st Wisconsin Cavalry, and called to active duty. It would serve on the Mexican border under General Pershing, a mission that lasted until mid-March 1917. Then Troop A was deactivated.

I always wanted the general to talk about his Mexican border experience. Patton served there also, although of course they didn't know each other at that time and Patton was a West Point graduate and a commissioned officer while Koch was an untried private. I didn't expect him to have come in contact with Patton, but I wondered if he had any insights on his future commander's performance in pursuit of Pancho Villa. We never got around to that discussion, though, as it seemed there always were more immediate things to talk about.

The deactivation of Troop A didn't last long. Soon after the United States entered World War I with a declaration of war on Germany in early April, the unit was called up again. But horse soldiers apparently were not the Army's greatest need, and the Wisconsin cavalry troop was transformed into a field artillery unit.

During his tour of duty in France, Oscar Koch was commissioned a second lieutenant in the National Guard back-dated to his twenty-first birthday. He was assigned to the famed French artillery school at Saumur as an instructor and saw action in areas of the country he would revisit in the next war. The second time around, his role would be significantly bigger. He would be G-2 of the U.S. Third Army, Patton's chief of staff for intelligence.

Lieutenant Koch came home to Milwaukee when the war ended. He'd seen combat and lived to tell about it. He felt good knowing he'd done something for his country and he had grown accustomed to the ways of the military. He enjoyed life in the Army.

"I knew it was the life for me," he said. "I had found a home."

He wasn't ready to become a civilian. This time, he took the initiative himself. He reorganized the old Milwaukee Troop A as the first federally recognized National Guard unit in Wisconsin and, a year later, resigned his Guard captaincy for a Regular Army commission as a cavalry officer. He bounced around from post to post, doing two tours at Fort Bliss, Texas, two at Fort Riley, and Signal Corps duty in places like Watertown, South Dakota, and Des Moines.

I was surprised to learn the general spent a brief tour in 1933 commanding a Civilian Conservation Corps unit, which he remembered as one of his most enjoyable assignments. He was in charge of some two hundred young men from southern Illinois who formed up at Jefferson Barracks, Missouri, and shipped to Washington State. That turned out to be a short-term location.

"We were up on Mt. Adams in Washington on the eighteenth of September and had an eighteen-inch snowfall," he recalled. "They mercifully moved us down to warm and sunny southern California. I always thought this kind of change in scenery was one of the real benefits of the CCC."

He said the CCC brought a starkly different lifestyle to some of these young men, "especially those from the larger communities like East St. Louis, and the work must have been pretty rugged for them for a while. But it turned out to be a good experience. They were good kids. They were happy to have a job and they worked hard. It made me feel good to work with them."

Years later, after he retired from the Army, he met some of his former CCC charges again in Carbondale and took some delight in recalling their good times.

While his Conservation Corps unit was in California, Koch met and became friends with Lieutenant Colonel Henry "Hap" Arnold, commander of a small airbase called March Field. Arnold later would become chief of the Army Air Corps.

For the next six years, which is a long assignment by military standards, Koch was an instructor in the Cavalry School at Fort Riley. He wrote articles for and edited the *Cavalry Journal*, and in 1935 the Army sent him to a summer session at the University of Michigan to study psychology and teaching methods. He wrote a thesis that would become required reading at many of the service schools.

Koch worked with Patton at Fort Riley. Patton had chosen the cavalry as his basic branch of service on graduation from the U.S. Military Academy in 1909 and was the Cavalry School executive officer. The two of them got on well and from here it was only a matter of time until Patton called Oscar Koch to join the 2nd Armored Division staff at Fort Benning.

"I couldn't have been happier than I was the day I got the 2nd Armored assignment," the general recalled, his eyes alight with pleasure as he recalled the occasion. "You knew Patton was masterful with armor. And even though we hated to admit it, the horse cavalry wasn't going to be fighting any more wars."

"So now you were in a combat unit with George S. Patton Jr. as your commander," I said. "You knew his reputation. Didn't this scare you a little bit?"

Oscar Koch chuckled. "If you were going to make it with Patton, you had to learn fast and do your job the way he wanted it done. There weren't any shortcuts."

He told about a personal experience to drive home his point. Although the incident wasn't funny at the time, he got a good laugh from telling the story. He and the division operations officer, or G–3, Lieutenant Colonel Robert W. Grow, were called into Patton's office one day and given a quick lesson on their commander's philosophy.

"I want the two of you to understand that I do not judge the efficiency of an officer by the calluses on his butt," Patton told them. They were then dismissed. The effect was immediate.

"We headed straight for the field. After that, no one had to remind us there were no desk jobs in Patton's outfits."

"He was a good communicator," I ventured.

"Yes, he could say a lot in a few words. And you couldn't miss his meaning."

Robert Grow went on to become commander of a crucial unit of Patton's U.S. Third Army in World War II, the 6th Armored Division. His senior aide, Lieutenant Cyrus R. Shockey, recalled that Grow was a great admirer of Patton and was "not bashful about adopting this style of leadership."[4]

So far as Oscar Koch was concerned, the Patton philosophy was a perfect fit with his approach to duty. A major reason for his success as Patton's G–2—aside from the fact the two of them complemented each other well in terms of temperament—was his hard work and thoroughness. Koch was masterful at collecting intelligence information, beginning with "what was happening on the ground," and interpreting the information at hand, all of which took immense effort.

Koch's work ethic gained the attention and the appreciation of the other Patton staff officers.

Robert S. Allen labels him "the sparkplug of Hq Third Army." Further, Allen writes in his 1947 Third Army history, *Lucky Forward*, given Koch's exceptional ability and unfailing effectiveness, Patton and the Third Army chief of staff often would assign him tasks that were outside his G–2 area of responsibilities.[5]

From my first discussion with the general about his long term of military service, I could see clearly that he loved his celebrated commander. He could hardly speak of "the Old Man" without tears forming in his eyes.

I learned, too, that he was reluctant to seek credit for his own extraordinary contributions for fear he might diminish the luster of the Patton name. The more time I spent with General Koch, the more my own respect for Patton grew. Surely a commander who chose officers like Oscar Koch for his staff was a commander on top of his game, and surely a commander held in such lofty respect by those who served under him had to be a remarkable leader.

Chapter 4

IN THE ARTICLE on General Koch I wrote for *The St. Louis Post–Dispatch*, I reported that he was "writing a book recounting in detail the techniques of combat intelligence as applied in the Patton commands during World War II." I said the manuscript was nearing completion and the general was seeking a publisher. This was true, but there was a great deal more to the story than I knew at the time. What I learned was not especially encouraging.

Shortly after the article appeared, John Allen explained that General Koch had struggled with his manuscript for some time and had not had much success in his search for a publisher. He believed the general had reached a point where he was open to working with a professional writer who might be able to help him rewrite and edit his material and possibly produce a manuscript acceptable for publication. He said the general wanted very much to see his book in print.

"Mind you, the project is dear to his heart," he cautioned. "He'll be very careful about who he works with. He's afraid his book will lose its originality in the wrong hands, and I can't blame him."

Then John Allen's eyes twinkled in a familiar way. This was how he looked when he had something to say that especially pleased him, usually a funny story or an "old saying" he found appropriate to the situation at hand. But this time he was serious. He said the general, given his satisfaction with the *Post–Dispatch* article, very likely would accept me as a collaborator. Would this be of interest?

I was surprised, but intrigued by the prospect. Even though I knew he was earnest, I said, "Mr. Allen, are you putting me on?"

"Just ask him."

I would, but the mere notion of such a query was intimidating. General Koch had a story that needed to be told, but I was afraid I'd sound presumptuous if I just asked directly to be his collaborator. I'd never written a book. My credentials were limited. Given his own stature, it seemed unlikely to me the general would consider someone with my limited qualifications to be his associate in an undertaking of this magnitude. He had far too much riding on the outcome.

I dragged my feet for a couple of weeks and then, on a day when I had occasion to visit the Koch home for some other reason, I brought up the topic, trying to sound casual, hoping to come across to the general as if I wasn't seriously proposing that he accept me as a collaborator. I made clear that John Allen had suggested it.

General Koch's response was not what I expected. He not only agreed, but he was enthusiastic. It was almost as if he had been waiting for me to ask. He would give me everything he had and see what I thought, and if I found something of value in what he'd done he was ready to go to work.

"Bob Allen always said there has to be a city editor," he

said. "That would be you."

I didn't know it at the time, but Robert S. Allen, the nationally prominent journalist and author, had been Third Army assistant G–2, working under Oscar Koch. The two of them still were close friends.

Given how quickly the general invited me to be his collaborator, I suspected after the fact that he and John Allen actually had discussed this more fully than I'd been led to believe. Whether they had or not, this was the beginning of the book, *G–2: Intelligence for Patton.*

And regardless of how it came about, I was extremely grateful for the consideration.

General Koch insisted from the outset that we would share equally in the credit. I said I didn't think we could do that. He was the authority; my role would be merely technical, helping to get the book ready for publication. He was firm: equal credit, "by Oscar W. Koch and Robert G. Hays." I finally won this point by arguing that editorial protocol called for *with* instead of *and*, which served to assure readers the book was his work. And surely this was appropriate. What could I tell anyone about combat intelligence? (Ironically, with *G–2* cited as a reference in scores of military history works in the years since its publication, the authors frequently are listed incorrectly as Oscar W. Koch and Robert G. Hays. I doubt that it matters.)

No one could be more agreeable to work with than Oscar Koch. But his modesty popped up quickly. He said there was no "big I" in the material he already had written, meaning the personal pronoun. Instead of referring to himself in the first person, he had used "G–2." This was the way he would like to do the book. Secondly, he wanted the use of Patton's name strictly limited. "I don't want anybody to think I'm capitalizing on the Old Man's name," he explained.

I thought both of these requirements would handicap our work, make it less interesting and less likely to be published. I understood his reluctance, but it seemed to me the Patton name was important in gaining the attention of both publishers and potential readers. I also wanted to use a chronological organization, as opposed to the general's tendency toward a topical emphasis. We could include the technical aspects of intelligence that he wanted to use, perhaps as situational examples.

He was talking about a text book on combat intelligence; my view was somewhat broader. On the other hand, this was *his* book and it should be written the way he wanted it.

The general was open to my thinking on every one of these issues. We might give it a go if I was willing, he said, and see how things worked out. He was fully confident that we could work together, that we would make a good team.

I was supposed to go back to work after we finished our discussion, but I was far too excited about what had just taken place. This would be a challenge. General Koch deserved the best work I could give him. But I was honored he had chosen me to work with and I couldn't wait to get home and tell Mary.

She was excited, too. When would we be starting work?

"I'm starting right now."

The truth was, I couldn't wait.

I came up with a few possible titles, each including the Patton name but downplaying it as General Koch wished. Each of them emphasized combat intelligence. He chose the one with "G–2" as the most prominent element. This was my favorite, as well, although I felt a bit guilty, knowing a publisher would find a way to highlight graphically the Patton name in the sub-title. Today, when I'm invited to speak about General Koch, I always carry a copy of the paperback edition of *G–2: Intelligence for Patton* to demonstrate this point. Patton's picture is on the cover, and in the title "Patton" is printed in red while the other words are black.[1]

I wanted to personalize the book more, which meant using the "big I" pronoun he had leaned over backward to avoid. If I wrote a sentence like "I ordered air photos," he'd say he would rather have it read "G–2 ordered air photos," or even the passive "Air photos were ordered." But he never was disagreeable, always respected my opinion, and in many instances deferred to my point of view without much discussion.

It seemed to me there was no way to separate Oscar Koch the man from Oscar Koch the intelligence officer, nor Oscar Koch the intelligence officer from Oscar Koch, Patton's intelligence officer. A chronological account of Koch's work as G–2 for the U.S. Seventh Army in Sicily and the U.S. Third Army across the European continent, Patton's major commands, was in order. Reporting the action in which Patton's forces were involved would put Koch's work into context.

Whatever the organization, I assured him, this would be a book about intelligence in combat. He seemed pleased.

From the time we began work on the *G–2* manuscript, I made it a point not to read books that related to the topic of military intelligence, General Patton, or even World War II except on a very broad scale. I didn't want to be influenced by other opinions or perhaps be taken in by erroneous information. Our work would be based entirely on material substantiated by Oscar Koch—most of it from his own astonishing record.

This decision looks good in hindsight, because much of what was in print at the time was wrong on things like the perceived intelligence failure preceding the Battle of the Bulge. General Koch's success in anticipating the German breakout was rarely reported accurately until after *G–2* was published.

There was one exception, and it was an important one.

I decided to read Robert S. Allen's *Lucky Forward*. I was familiar with Allen's work as a journalist, in particular the very popular "Washington Merry–Go–Round" syndicated column he co–wrote with Drew Pearson. Allen and the general corresponded regularly, sent each other written material to critique, and shared stories and gossip about old associates. Allen's book was hard to find—it may have been out of print—but readily available in the university library. (It actually was only a few years ago that Mary was able to get a personal copy for me through our old favorite, the wonderful Jane Addams Book Shop in downtown Champaign, Illinois.)

Nan Koch let it be known, albeit subtly, that she was not particularly happy with *Lucky Forward* and told me General Patton's widow, Beatrice, didn't like it at all. After reading the book, I thought I knew why. I suspected Beatrice Patton felt that Allen gave Patton's staff too much credit, and I was virtually certain that Nan Koch felt he didn't give Oscar enough.

If my hunch was correct, I would not have blamed Beatrice Patton too much. In Nan's case, though, after working with

General Koch on *G–2*, I had a feeling he may have had a role in the limited use of his name. In his first reference to Oscar Koch, Allen was fulsome in his praise. He said flatly that "Koch is the greatest G–2 in the U.S. Army. His record is without equal in every phase of intelligence."[2] But through the remainder of the book, he seldom mentioned Koch by name.

Instead, he used a generic reference, "G–2" or "the G–2 Section." Sound familiar? Yes, this is exactly the way General Koch wanted us to do it in *G–2: Intelligence for Patton*. I have little doubt he made the same request of Robert S. Allen.

These assumptions were borne out, at least partially, when Hirshson's Patton biography came out in 2002. But there is a twist I didn't expect.

Hirshson writes that Beatrice Patton was unhappy with Allen's book in part because she believed he had "practically plagiarized" Patton's *War as I Knew It* (Allen had seen a pre–publication copy of the manuscript) and in part because she felt Allen didn't give Oscar Koch enough credit.[3]

Hirshson says Mrs. Patton complained that *Lucky Forward* might very well leave a reader with the impression that Allen was Third Army G–2 instead of Oscar Koch. This also was my reaction to Allen's otherwise fine work, which is precisely why I reasoned that Nan's complaint was based on the protective attitude toward her husband's name and reputation that I witnessed on any number of occasions.

Beatrice sent Oscar Koch a copy of Patton's *War as I Knew It*. She inscribed it, "To the Greatest G–2."

Once the general and I began work in earnest, I learned

quickly that his writing efforts went back a great deal farther than I'd realized. He actually had completed a manuscript titled *Intelligence in Combat* as early as 1955, although he'd made no effort toward publication until he'd received clearance from the Department of Defense. This came in late 1956. Frederick A. Praeger, the New York publisher whose specializations included military science, already had shown interest.

Praeger's publications list was impressive. It included some of the books of the noted British military historian, Captain B. Liddell Hart. But Praeger eventually rejected Koch's manuscript, citing an uncertain market for what essentially was a textbook on combat intelligence.

Meanwhile, General Koch had been in touch with other publishers. Over time his manuscript was considered by McGraw–Hill, Vanguard Press (Robert S. Allen's publisher), Alfred A. Knopf, the Cornell University Press, and Macmillan, among others. All were complimentary about his work, but in the end all decided the book was too specialized and would not have a large enough market to make its publication profitable.

In some instances, individual editors who favored publication only to be overruled by higher authority in their own respective publishing houses suggested other possible outlets and named individual contacts. An editor at Macmillan said he thought his company had made a wrong decision in turning down Koch's book. The editor insisted that he, personally, still believed in its potential. He recommended McDowell–Obolensky and, with Koch's permission, sent the manuscript there. It was once again rejected.[4]

Macmillan sent the general a copy of an Army War College reviewer's report which was highly complementary. The reviewer, whose name was not revealed, said *Intelligence in Combat* would contribute materially to the understanding of its subject not only by military personnel but by civilians as well.

And he added, "Having read the manuscript, I feel that my professional competency has been increased."5

In May 1960, Koch scheduled a luncheon meeting at the Pentagon with Major General Robert W. Porter Jr. to discuss an entirely unrelated matter, but Porter expressed interest in his book. It turned out that he was a member of the Chief of Staff's reading panel which made recommendations for reading lists at the various service schools. General Porter had served as 1st Infantry Division G–2 in Tunisia and after that as a corps G–2. He asked to see the manuscript, and later reported that he definitely would recommend the book for required reading at the service schools. But first, it would have to be published.

General Koch visited Guggenheim Foundation president Henry Allen Moe in New York and brought him up to date on all his contacts with publishers. He told Moe he'd misjudged the value of a book on intelligence techniques, and felt he owed it to the foundation to rewrite it with a more personal flavor to give it broader appeal and increase the chances for publication. The Foundation head was not certain this was such a good idea.

"Mr. Moe asked why it had been written as it is, and I told him that that was the way I felt it would be of greatest future benefit," the general wrote in a memorandum on the visit. Moe told him if this was the way he felt, it would be a mistake to rewrite it. The organization's president also said the foundation would be willing to help underwrite publication costs if that might prove useful.

A possible publishing subsidy attracted the attention of Stackpole, the former Military Services Publishing Company. Once they had estimated costs, however, they set the amount required much higher than the foundation was willing to pay.

Meanwhile, General Koch and Henry Allen Moe discovered they both planned to be visiting relatives in Carbondale, Illinois, over the Christmas holidays. Although they would be in Carbondale at different times, Koch arranged to drop off a copy of the *Intelligence in Combat* manuscript at the home of Moe's son so that Moe could read it during his visit and leave it for the general to pick up later.

In January, Moe wrote that he'd "found your manuscript fascinating: I learned much and I enjoyed the learning. And I hope that, not only for the benefit of military men but also for the public, your manuscript will obtain adequate publication."[6]

General Koch's book subsequently was rejected by Doubleday and Prentice–Hall, the latter suggesting it was "a superb textbook" but too specialized for them. In what must have felt like a slap in the face to the general, Prentice–Hall also reported that in response to an extensive market survey a lieutenant colonel who taught combat intelligence in the Tactics Department at West Point had replied, somewhat surprisingly, that "the Academy devotes little class time to the subject and does not require any text material."[7]

The general had become discouraged and almost given up looking for a publisher. Then, in early 1963, he was contacted by Ladislas Farago, who said he was working on a book about Patton and the U.S. Third Army for Random House and would like to ask General Koch some questions. He said he felt Third Army had not received proper credit for its great accomplishments and Robert Allen's *Lucky Forward* needed to be updated.

Farago came to the Washington, D.C., home of General Koch and Nan the next morning and spent several hours there. He presented the general with a copy of his most recent book, *The Tenth Fleet*, and spoke of other former Patton staff members he'd been talking with.

Farago's first question concerned an incident he said was reported in Charles R. Codman's book, *Drive*. He wanted to know more about the day Koch woke Patton at 4 a.m. in connection with the beginning of the Battle of the Bulge.

General Koch told him he was confused; the incident in question was not related to the Bulge, but rather to the encirclement of the German Seventh Army in the Falaise–Argentan area in the aftermath of the Allies' invasion of France.

The visitor wanted more information. Koch said the simplest way to get it would be to read his manuscript. Which manuscript? Farago asked.

When the general told him about *Intelligence in Combat*, Farago showed great interest. Why had it not been published? No market, the general said. Could he see it? The general said yes, but stipulated that if Farago used anything from it in a book, he expected full credit.

What happened after that is detailed in Oscar Koch's *aide mémoire*, an example of the meticulous notes he habitually kept on important events:

From that time on he was more interested in the MSS than his questions. We talked about 3 hours—with him doing some reading at my suggestion, and he said it was a masterpiece.

Nothing amateurish, entirely professional, and with a clear ring of authenticity. It was a book that HAD to be published. Not only that, but he had a publisher in mind. He said he could handle it two ways: one, where he would have to argue to have it published, and the other where he could say he wants it published and that would be that. It all looked almost phoney [sic], except that he knew nothing of me or my MSS, and was intensely interested.[8]

General Koch was rather favorably impressed with Farago, who talked about his experiences in Naval Intelligence in World War II and his association with Radio Free Europe. The visitor also had an impressive list of publications to his credit. He told the general he could make an offer on the spot. Would Koch accept $2,000 against royalties, under contract? The general said yes.

Would he object to an immediate paperback? The general said he would not.

General Koch also advised Farago that he was not willing to do a major rewrite on the manuscript and if a publisher took it, it would have to be on an "as is" basis. He apologized for "not being too excited" about the offer. He had been close to publication before, only to be disillusioned by having the deal fall apart later. Farago said the two publishers he had in mind were Random House, where he would have to argue for publication, and Obolensky, where he simply needed to tell them he wanted it published and it would be.

The general's *aide mémoire* continues:

> I never got into his relationship with Obolensky but told him that they had had the manuscript and had turned it down a matter of a few years ago... He commented that the firm had changed... and that in all probability it found its way into the hands of some stupid editor, who couldn't see its potential, which he, Farago, could instantly recognize. How best to handle it from here on? I suggested that I would appreciate a letter from New York asking for it and he said he was returning to New York the next day, and that I would hear from them. That I did, and Friday afternoon, the MSS was on the way.[9]

General Koch immediately notified his friend Robert S. Allen. "Bob Allen is thrilled," he would write later. Allen agreed to "check out" both Farago and the Obolensky publishing firm in New York and asked if he might look over the contract before the general signed it to make sure it covered everything that Allen considered important.

The general promptly received a telegram from Ivan Obolensky, Inc., the New York publishing house, acknowledging receipt of his manuscript, and then further confirmation in the form of a post card. He also got a call from someone in the Obolensky office who said Farago was standing alongside and the publisher's only concern was when the work might get into print. It was too late to get it onto Obolensky's spring book list, he said, but it would be on the fall list.

About a week later General Koch received a letter from Farago. The writer thanked him warmly for his hospitality.

With regard to the *Intelligence in Combat* manuscript, Farago reiterated his full confidence that "publication is assured." He told the general, "I am very much impressed with your writing and while the book may need some elaboration here and there, its structure and presentation need no fundamental alterations."

Farago then went on to offer the general "My heartfelt congratulations!"[10]

Having heard nothing more from the publishers a month later, the general sent a polite query. He said he planned to be out of Washington for four to six weeks "unless the manuscript requires more immediate attention in the meantime." He got no response. After yet another six weeks, he wrote again. He said Farago had intimated that a contract could be "expected forthwith." Unless the manuscript was being favorably considered for publication, he wanted it to be returned without delay.

This brought a phone call from New York. A representative of Obolensky arranged a visit to the general in Washington. General Koch's *aide mémorie* relates what happened in that four–hour meeting:

> The contract was gone into in great detail, from now on we'd talk about the book instead of the MSS; the book should be in two parts, one the stories, and the other the techniques, the title might be changed to what? They were

interested in putting out my book as a companion to Farago's new one: two for the price of one sort of a deal, both about Patton, both about WWII, both about techniques, one of command the other about staff, etc.[11]

The visitor promised Oscar Koch he would get a letter of intent within a week.

After another six weeks had passed during which he'd heard nothing, General Koch wrote the publishing house again, once more asking for an update. A month after that he sent a registered letter to the publisher demanding return of the manuscript. And yet a month later, after placing a number of phone calls, he finally talked with Ivan Obolensky, himself.

General Koch outlined to the publisher the long chain of events relative to his manuscript.

Obolensky said the firm's representative the general had met with had asked to handle the book personally and he, Obolensky, was surprised to hear there had been a problem. The general informed the publisher that he would not enter into any agreement with the firm after the "shameful" treatment he had received and Obolensky promised to return his manuscript as soon as he could find it.

The general received a letter from Obolensky dated two days later. "I must say that while I am captain of the ship, I accept full responsibility for the lack of follow-through by my staff," the publisher wrote. "As far as I am concerned, I personally reviewed your manuscript some time ago and my decision at that time, directed to one of my employees, was to enter immediately into contractual relationship with you. It never got any further, apparently, and your call was the first that I had heard of the matter since I had given it my full prior approval."[12] He then wished the general the best of luck in getting his book published.

Oscar Koch was too polite to reply in the manner he surely wanted to, or certainly in the manner I would have. He wrote Obolensky a short, formal letter thanking him for "your kind and courteous letter." But he did complain about "the business procedures to which I was subjected."

Ladislas Farago, whose book, *Patton: Ordeal and Triumph*, was published by Obolensky in 1964, personally returned General Koch's manuscript a week later.

Perhaps the general's reward comes through his characterization in Farago's book. Farago describes Koch as Patton's "meticulous intelligence chief" and says he deserved the Medal of Honor for his courage in August 1944 when he stood up against the over–optimism pervading the top Allied command at SHAEF (Supreme Headquarters, Allied Expeditionary Forces) that the German Army was finished—a belief that in large measure led to Eisenhower's and Bradley's failure to respond in advance to the enemy buildup before the breakout in the Battle of the Bulge.

He points out that Koch's intelligence estimate on August 28 leaned hard in the opposite direction. Oscar Koch refused to concede that the Germans had been decisively defeated and, in spite of heavy losses inflicted by the Allies, he said the enemy had been able to maintain "a sufficiently cohesive front to exercise an over–all control of his tactical situation." Koch flatly refuted Eisenhower's G–2, General Kenneth Strong, and said it could be expected that the German armies would "continue to fight until destroyed or captured."[13]

So far as the story of Koch's *Intelligence in Combat* is concerned, there was one final chapter. Not long after he and Nan moved to Carbondale, the Southern Illinois University Press expressed an interest in the general's manuscript. After he delivered it, the press sent it to a reader for evaluation.

The reader's report not only was negative, but in fact truly scathing in its criticism.

Press director Vernon Sternberg suggested to the general that he consider getting a professional writer to help rewrite the manuscript and referred him to a writer in Connecticut.

General Koch's letter to that individual effectively illustrates his remarkable ability to maintain his sense of humor even in the face of trying circumstances, in this case a reviewer who apparently went overboard in his criticism. He told the professional writer the net value of his manuscript had now been established as lying somewhere between the extremes of being worthy of required reading at the military service schools and, by the latest reader, as "incredibly poor" and "unfit for publication because of a style which is both prolix and puerile, and at times unintelligible."[14]

Mercifully, the Connecticut writer replied that the general ought not to be dismayed by the reader's remarks. "For some reason, which I cannot understand, the reading of manuscripts brings out the savagery latent in man and woman, and they can rise to heights of fury that seem to be incredible over a misplaced bit of syntax."[15] The writer specified a fee, and the general wrote back that "lacking assurance of publication" he'd decided not to commit himself to added expenditures.

Oscar Koch's treatment by a number of publishers is hard to comprehend. His *Intelligence in Combat* manuscript was worthy of their attention. On the other hand, I am happy that when my turn came the opportunity for collaboration still was open.

Chapter 5

There is little doubt I would have grasped the exceptional level of General Koch's performance as Patton's intelligence chief during World War II more quickly if I had been an adult during that period of American history, and especially if I had been in uniform at the time.

But I was still a child when fighting in Europe ended with General Alfred Jodl's signature on a simple, 234-word typewritten document of surrender at Reims, France, on May 7, 1945. Jodl signed on behalf of the German High Command, while General Walter B. ("Beetle") Smith, General Dwight Eisenhower's chief of staff, and Russian General Ivan Sousloparov signed for the Allies.[1]

Even as a child, though, I had understood very well that my country was at war.

The grownups listened to the radio for news from the front lines and read it in their newspapers and magazines. The newsreels we watched in the movie theaters offered a more graphic view, sometimes painful to watch.

I remember the blue service stars in the windows of homes from which a family member had gone to serve, and we knew what it meant when the blue stars were replaced by the gold.

There always was talk about someone who had lost a loved one—a son, husband, brother, father, uncle—in a far-away place like Italy or a remote island in the Pacific. I knew who was on our side and who our enemies were. Nazis and "Japs" were held in high contempt.

We listened to Eddie Cantor's recording of "A Wing and a Prayer" and understood that it was a paean to the courage of our pilots in the war and an expression of hope for their safety.

There are pleasant memories, also. In school, we collected empty tin cans and picked up any random scraps of metal we could find for the war effort. We gathered the down of dried milkweed pods to be used as filler in life preservers which we hoped would save our sailors if their ships were torpedoed and our airmen if they crashed at sea. We felt good to be doing our part.

Walnut Grove, my one-room country school in southern Illinois, sat alongside a dusty gravel road amid tall white oak trees. Yucca plants—we called them "needle and thread"—and tawny lilies grew just outside the front fence and there was a long-handled iron pump we used to draw up water from a shallow well. Playground apparatus was a set of swings and a slide. Most important, there was a level dirt basketball court. On cold mornings we often began playing basketball when the ground was frozen and kept on until the court surface thawed and turned into a sea of thin mud that coated our shoes and dirtied our hands and splashed on the legs of our trousers up to our knees.[2]

Our school was some thirty-five miles west of Evansville, Indiana, where a sprawling Republic Aviation factory built P–47 Thunderbolt fighter planes. There were days when formations of Thunderbolts flew over, getting run through their paces by test pilots—we know now that many of them were women—and roared through maneuvers like power dives or steep climbs at full throttle. Our teacher seldom fought the competition; he'd let us go to the windows and strain our necks trying to see the action, or if we were lucky and he was in an especially good mood he would let us go outside and watch.

There were large flights of bombers, too, and if they were at high altitude I watched them from horizon to horizon. I was fascinated by their formations and wondered where they were headed, whether they would soon be in action. Sometimes the flights were low. Late one hot summer afternoon, six B–25s flew over our house just above tree–top high.

Our teacher read aloud to us for a half–hour each day during quiet time after lunch. One of the books he chose was *Thirty Seconds over Tokyo*, Captain Ted W. Lawson's first–hand account of Lieutenant Colonel James H. Doolittle's carrier–launched bombing raid on Japan in April 1942.[3] The book obviously could not portray the drama of Doolittle's sixteen fully loaded B–25 bombers taking off from a carrier deck the way the later movie version did, but it offered a marvelous story that kept a roomful of school children eager for the next installment.

When I saw the movie some years later, there were scenes that reminded me of the low–flying B–25s I saw that day, standing in my own front yard, awed by the speed of those marvelous airplanes and shaken in my tracks by the noise of their powerful motors.

Even if I didn't truly comprehend the cost of war, I was happy to hear that this one was over. Now we could get more sugar and gasoline and meat, and tires for the rickety bicycle my brother and I had cobbled together with parts from various older two-wheelers. And my Uncle Buck would be coming home from England. I couldn't remember when he left, only that he'd been in the Army for a long time.

It wasn't until I was in high school that I came to understand the true horrors of the conflict my country had been through. In civics classes, homeroom, even in full student body assemblies, we saw movie after movie depicting combat, destruction, Allied prisoners being rescued from German stalags, flame-throwers burning Japanese soldiers out of their caves on Pacific islands no one ever had heard of, the hostages of Hitler's horrible concentration camps as they were liberated by American soldiers. And I learned a bit about a flamboyant American general, George S. Patton Jr.

I suppose it was hoped that if we saw what happened we would never forget. If so, I believe the effort worked the way it was intended. Those actual scenes of terrible warfare left an indelible impression on my teenaged mind, so much so that it is hard for me even to imagine the effects that actually being involved in combat would have on anyone. And it is similarly difficult for me to imagine Oscar Koch, the gentle man we knew, having been at his best in that awful field of death and destruction.

At 18, I probably never thought about military service. The Korean War came and went and this time I had friends not so much older than I who lost their lives. And even after that war ended, the draft still hung over the heads of young men so that their futures were uncertain. My brother was drafted. I couldn't afford college and jobs were hard to come by. Shortly after I turned 20, I volunteered to have my name put at the top of the draft list. A few weeks later I filled one half of that month's conscription quota for White County, Illinois.

Military decisions come from the top down. But draft boards were local. When I answered the call to mine, I found a familiar face there.

Walter B. Young Sr., representing the White County Draft Board, had been our rural mail carrier for as long as I could remember. I had waited beside our mailbox many times, on a mossy bank in the shade of a misshapen old oak tree, watching patiently for his ancient Ford or his "new" 1941 Chevrolet. In most instances I had cash in hand to buy stamps or maybe a postal money order to stick in with my mother's shopping list from the Sears-Roebuck catalog. Sometimes I simply was hoping he'd deliver one of the box holders with free samples of chewing gum that came frequently back then.

On this day at the draft board office he gave me a silver dollar.

"I know you won't be in combat," he said, "but keep this with you for good luck. Be safe, lad, and come back to us after your job is done." There were tears in his eyes.

It is ironic to me that at the very time my brief hitch in the armed forces commenced, the long and distinguished military service of General Oscar Koch had just drawn to a close. I do not believe in manifest destiny. General Koch and I were far apart in age and, on the surface at least, had little in common. Surely the odds of our meeting were not great. But we live in a world of chance, and fate is an inescapable force at work in every human life.

Oscar Koch and I were linked for a few hours through mere coincidence, and that few hours led to a few days and then to months and years. His influence on me has lasted for a lifetime.

My military record was insignificant, but the experience gave the general and me a point of common interest and things to talk about. Clearly it helped me to better understand his lifelong commitment and dedication to duty. As much as anyone I knew, he was proud of me for having been under arms. He gave me credit for accepting the discipline of military life without complaint. This was a way of living he'd chosen for himself, and for which he had no regrets, but he understood that not everyone is suited for the regimented Army life.

In my case, I knew beforehand that I was one of those who would not enjoy the limitations imposed by the military. But I was not prepared for how quickly I would experience those limitations. Once I was in the government's grasp, I learned immediately how little I had to say about what happened next.

This extended even to the branch of service in which I was to serve. It simply was my place in line at the U.S. Fifth Army area induction center in St. Louis that landed me in the Army rather than the Navy. I was sworn in, along with perhaps three dozen other men, and then we were lined up and marched through an open door. On the other side stood an Army officer and a Navy officer, one on the right and one on the left.

Each tapped every other man on the shoulder as he came through the door: "You're Army." "You're Navy."

The rest of my life would have been different had I been one step ahead or one step behind the position I held in that long human queue. But as I said, we live in a world of chance.

Military experience in itself leaves vivid and lasting memories. Even if it doesn't lead to combat, the intensive training forces total commitment, along with a sudden and dramatic change in lifestyle and social surroundings. I went through basic combat infantry training at Fort Leonard Wood, Missouri, in November, December, and January 1955–56. Decades later, as I wrote my fourth novel, *Blood on the Roses*, recollections of that experience still were as fresh as if it had been only months past. One of the principal characters in my novel is Sergeant Willie Jamison and my personal memories of Fort Leonard Wood were the basis for an important part of his life story.[4]

General Koch never told me much about his early years in the horse cavalry, but I was left with the impression that most of his training was in the form of mounted drill. Nothing really comparable to the combat infantry training developed later. Few of the physical rigors. On the other hand, I didn't have to worry about learning the proper way to ride and I didn't have to take care of a horse.

And no disrespect to the general or any other old horse cavalryman, but a mounted charge waving a Patton Saber does not strike me as the way I'd prefer to go into battle.

Combat infantry training is designed to develop physically fit soldiers in a limited period of time. I was by no means a sterling example of fitness. At just over six feet tall and weighing only about 140 pounds, I was a skinny kid. But I was a country boy accustomed to the hard work of farm labor and construction jobs in the southeastern Illinois oil fields. From third grade on I'd played basketball and run track in school. Basic training was not fun and games, but the physical challenges were not that tough.

Fort Leonard Wood was home to the 6th Armored Division, reactivated in 1950 as a replacement training division. I knew nothing of the history of the "Super Sixth" and its role as a unit of Patton's U.S. Third Army in World War II, but I rather fancied the colorful, triangle–shaped division patch I wore. The division was deactivated again in the spring of 1956 and Fort Leonard Wood became the home of the United States Army Training Center, Engineer.

The Army training methods were effective.

Instructors took nothing for granted. Everything started at the beginning, assuming we trainees knew nothing, and proceeded systematically, step–by–step. "By the numbers" is more than a cliché. It is the way the Army taught us how to fight, to use the weapons of war.

As opposed to the teaching methods, though, the settings chosen for our instruction sometimes left much to be desired. On bitter winter mornings, before daylight, we often were marched to far–flung training areas miles from our barracks, dressed in full winter gear, our breath making icicles on our noses and chins, only to sit on cold outdoor bleachers and be subjected to boring fifty–minute demonstrations on how to clean the M–1 rifle or whatever. Hardly the best classroom for learning. We were much more interested in the next break, when we could jostle for spots nearer the fire barrels, than we were in cleaning rifles.

But there was one outdoor training event that impressed me. One night my unit was taken out into the woods to observe a firepower demonstration. We sat on the side of a steep bluff and watched and listened while training cadre in a staged encampment on the opposite side of a deep ravine, perhaps two hundred yards away, filled the darkness with talk and laughter. When one of them lit a cigarette, it was as if a flare had been ignited.

That was the prearranged signal for our carefully orchestrated enlightenment to begin.

"When you're in enemy territory," our instructor bawled, "you keep your mouth shut and your head down. Those guys just cut their lives a whole lot shorter."

Then, in the truly dramatic demonstration we had been brought there to see, a single rifle squad arrayed along a trench before us opened fire. One squad member had a Browning automatic rifle, commonly called a BAR. All the others were armed with M-1 rifles. They all used tracer ammunition so we could see the bullets in the pitch-black night and they sprayed the area across the ravine where the "enemy" troops ostensibly were resting, unaware that they had been observed.

It appeared to me that, had this been actual warfare, none of those men could have survived. Surely this single rifle squad, armed only with the weapons they carried, would have obliterated all life in that compact target zone. This one simple demonstration left me with tremendous respect for the firepower of a single rifle squad engaged in battle and the losses it could inflict on an enemy.

But like the young Oscar Koch, I did not expect to see much real action. I never felt that one day my life might depend on what I learned at Fort Leonard Wood during the frigid winter months of 1955–56. And the Army apparently agreed. If I happened to draw KP duty on a day that caused me to miss firing the machine gun or throwing grenades, it seemed as if firing the machine gun or throwing grenades wasn't too important. At no point was I ever required to go back and make up something I'd missed because of KP or some other work detail.

Fort Leonard Wood is situated at the northern edge of the Ozark Mountain range, and it is a bleak place in the wintertime. It provides terrain in many regards ideal for combat infantry training.

The only event of note during the months I spent there was a ceremony retiring the colors of the 6th Armored Division, in which a great mass of trainees and post cadre marched in a long parade before a reviewing stand populated by a handful of people. I suppose there were dignitaries among them and I assume we were reviewed by the commanding general, but given Fort Leonard Wood's location it didn't seem likely to me that any ceremony there ever had drawn much of a crowd.

Trainees, though, know little about what goes on around them. Except for those in my own company, the only Fort Leonard Wood officer I remember hearing much about was the deputy post commander, General Normando A. Costello.

General Costello had a reputation for driving around the base looking for things that needed to be spruced up. Unit commanders hated to hear that he was on the prowl anywhere near their area. I learned years later that General Costello had been a close friend of General Oscar Koch, but I never knew the connection.

After basic training, I stayed at Fort Leonard Wood. I merely moved a couple of blocks to a different company for advanced instruction in what we generally called "clerk-typing school."

No heavy lifting now. Training here was in an indoor classroom that might have been in any secondary school in the country. We drilled on rickety Army typewriters and learned the elementary rules of Army administration. Because I had taken a journalism class in high school and decided I wanted to be a reporter, I'd also taken a class in typing. My keyboard skill was not impressive by any means, but it was good enough to give me a modest head start.

My new company was adjacent to a unit that trained combat engineers. We felt lucky when we finished classes at five o'clock and went back to our warm barracks for the night. It seemed as if the engineers always were out in the frigid darkness, busy building a bridge or stringing communications wire or the like. We would make bets among ourselves on how long it would be before they were back to dismantle whatever they'd just built. They usually were back again the next night.

General Koch would have appreciated my account of the engineers' activity. He knew about stringing communications wire. When he was signal officer for the 14th Cavalry at Fort Des Moines, he directed the installation of the first communications line between Fort Des Moines and Fort Crook, Nebraska. But this is another of those small things we never had time to talk about. I'd love to be able to sit down with him now and discuss it. He'd have a good story or two and we'd have a good laugh, and he would chide me over my "soft" duty at the time.

Except for the occasional guard duty, we budding "Remington Raiders" spent those cold Ozark nights in relative comfort. One of the guys on our floor of the barracks had a portable record player and one album. Every night we went to sleep to the soothing music of Bobby Hackett's magic trumpet.

Administrative school passed quickly. Many of the troops

in every graduating class, regardless of their field of training, were being sent to Korea. The war there was over, but Korea still posed harsh and dangerous conditions and very few soldiers actually wanted that assignment. I felt good when orders were posted and I and three of my friends were transferred to Fort Jackson, South Carolina. Little did I dream of the effect this posting would have on the rest of my life.

Chapter 6

SHORTLY AFTER GENERAL Koch and I began work on *G–2: Intelligence for Patton*, we learned that our mutual close friend John W. Allen was seriously ill.

Aside from the fact that we were immensely indebted to him for his role in getting us together on the book project, we both thought very highly of him and we were deeply concerned. He was in his early eighties and we knew his prospects were not good.

The general visited with him regularly. It soon was apparent that the bond between them was even stronger than I'd realized.

The two men had common experiences that surely gave them much to talk about. Like the general, John Allen fought in France in World War I. And like the general, he entered service as a volunteer. In his case, it was with the U.S. Marine Corps. Only a few months before his final illness, he had visited France and retraced his steps on the battlefields where he'd been in combat decades earlier.

They also shared the experience of reaching remarkable levels of success with only limited formal education. The general had graduated from high school in Milwaukee before joining the Light Horse Squadron Association, but John Allen, even though he was to become a prominent educator and historian, lacked even a high school education.

He passed a required teacher's examination after completing the eighth grade at a one-room country school aptly named Hardscrabble. (I had heard him tell stories about Hardscrabble when I was a student, but I did not realize at the time it was an actual school.)

Another thing they had in common was their reticence to talk about themselves. Irving Dilliard, the Pulitzer Prize-winning former editorial page editor of the *St. Louis Post-Dispatch*, wrote in his foreword to John Allen's book, *Legends and Lore of Southern Illinois*, that he would make good use of his space "to tell things about John Allen which he would balk at telling about himself." He went on, "For our author is a modest man. In all this book, the reader will find almost no reference to John Allen's origin, to his background, his parentage, his family, his personal experiences—where he has been, what he has seen, why he has lived the kind of life he has."[1]

I recently had written a feature story on John Allen for the Volkswagen company's *SmallWorld* magazine, a publication sent out to owners of Volkswagen automobiles in the U.S. The editors had titled the article, appropriately, "Teller of Tales." In reference to his old age, I had quoted what he characterized as being part of the John Allen philosophy, "I want to work—easy, of course—till sunset."[2]

John Allen soon passed away, and Oscar Koch was called on to deliver his eulogy. He talked about his old friend's rich life and many accomplishments. Then he, too, touched on John Allen's modesty. As the general explained it, "It took years to find out about the honors he had received, the recognition he had been given, and the romantic and rustic life he had led." General Koch described their time together:

> His space in the Morris Library building [at Southern Illinois University] became the scene of almost daily conferences and although we had agreed that there would be little or no time wasted reminiscing of the days of yore—many of which we shared in common interest—there were many pleasant hours spent talking about his native and my adopted southern Illinois.[3]

Losing his friend was difficult for the general. They had become extremely close as key leaders of the Carbondale Memorial Day Association, and John Allen had led him to become active in the Jackson County Historical Society. The only time I ever heard General Koch voice even mild disagreement with Nan was once when she expressed gentle criticism of John Allen.

He put his disagreement softly: "Maybe. But nobody asked what I think."

It seemed as if the general's bereavement ignited within him a renewed determination to get on with the things he still wanted to do in his own life. Specifically, this meant getting his book into print. Or perhaps he merely grew more intent on this project as a way to keep busy and not have time to grieve. In any case, he was eager to meet with me more frequently, more enthusiastic about seeing material I'd written or edited, more vocal about keeping our efforts going.

This meant, for me, plunging more deeply into the record of the general's work. He could update some of the best of what he'd already written, especially background matter on the history of military intelligence as it pertained to the American armed forces. I would focus on his work during World War II. The starting point was North Africa.

The more I learned about the planning for that 1942 seaborne invasion, the more I admired those who faced that formidable responsibility, who went about the task with equanimity, who accepted the fact that when their nation was at war they had a duty to fulfill. The lives of thousands of troops who would take part in the landing were in their hands and the course of the war would be vitally affected by the outcome of their operations. What kind of men were these?

The general brushed aside my question with a simple, "Just soldiers doing our job."

Yes, but surely the job was an immense challenge?

"It was a challenge, yes, but we knew what we had to do. You can't overlook anything."

"For want of a nail?" I said.

"Yes, for want of a nail a shoe was lost, for want of a shoe a horse was lost. You know the story. The biggest worry in planning for a military action of any size is that you might overlook some small detail that turns out later to have a big impact. Once the action begins, you don't have the luxury of going back to do it over."

The general was answering my question. What kind of men were these? The kind of men who understood the ancient parable about the loss of a nail leading to the loss of a kingdom, who recognized that the effects of a small and innocent mistake might be measured in the number of lives lost. They were the kind of men who worked as many hours a day as their minds and bodies would tolerate and devoted themselves to doing whatever it took to assure that American forces were ready for the battles that lay ahead.

The intelligence work preceding the North African invasion was so detailed the individual soldier would know exactly where he was going, what he could expect to face when he got there, and how he fit into the larger picture relative to the thousands of other troops involved in the operation. The massive invasion fleet's combat-loaded ships (meaning "first off, last on") carried sixty tons of maps to be distributed before the actual landing.

"And you must remember," the general said, "all the planning took place back in North Carolina, at Fort Bragg."

"Even the intelligence work?"

"Especially the intelligence work."

He went on to offer a straightforward and effective example of the value of pre–invasion intelligence gathering. Patton's objective was Casablanca. General Harmon's sub–taskforce Blackstone, of which Koch was chief of staff, was to hit the beach on the Atlantic coast near Safi, French Morocco. Once they gained control of the Safi area, Harmon's troops were to secure a crossing of the Rbia River so that Blackstone's tanks could be used to attack Casablanca from the south.

Invasion planners had requested and received air photo reconnaissance missions over the Rbia from the British Royal Air Force. With the advantage of ample aerial coverage, they knew what to expect when they got to Morocco.

The main resistance faced by Blackstone collapsed

quickly, as the French had little desire to fight the American forces. The French and American officers soon were meeting socially and exchanging toasts to Franco-American friendship and cooperation and declaring optimistic pledges to bring down Hitler and Mussolini. General Harmon hosted a dinner for commanders who only hours before had faced each other as enemies.

"One of the French officers was the commander of an infantry battalion that had been charged with defense of the main highway to Casablanca," General Koch recalled. "He said if he'd blown the Rbia River bridge we would have been delayed indefinitely." Not so, Koch responded. Had the defending troops blown the bridge, Harmon's forces were ready to throw a pontoon bridge across the Rbia 17 kilometers upstream. And failing that, they could have crossed a powerhouse dam another 13 kilometers farther up the river. And there was more.

"At the site where we were prepared to cross," he added, "the Rbia is only about 215 feet wide, with good approaches on both sides. It should be at normal stage this time of year, but just in case the water was high, we had brought an extra fifteen feet of bridging."

The French commander was astonished. How did the Americans know all this?

"Our intelligence had been at work," Koch told him bluntly.

I suggested to the general that we should use this story in our book, given that it illustrated very well the role of intelligence in planning a mission. I was surprised to see he was hesitant. I should not have been.

"It sounds like I'm bragging," he said.

"But you weren't G–2 of Blackstone, right?"

"Yes."

"Then you'd be bragging about somebody else's work, would you not?"

"Good point," he said, with a big smile. "Let's use it."

I wrote nearly a page of text for *G–2: Intelligence for Patton*, based on his story about the Rbia River bridge. I worried that the general might have second thoughts and perhaps overrule me once he'd seen the typescript. I should have known better. That wasn't his way. He'd already voiced his approval and he accepted the material without question. Inconsistency was never a tag that could be placed on Oscar Koch.

A smile spread across his face as he scanned my copy. His only comment was, "It looks too good." He meant the quality of the typewritten page, not the content. This was the first copy he'd seen that had been typed on a new IBM electric typewriter I had just gained at the office.

There still was a wealth of information about the campaign in North Africa to work with. After the shooting stopped in French Morocco, Patton's forces were reorganized to meet changing circumstances. The Western Task Force became the I Armored Corps and Patton's staff began looking to the east, where the Afrika Korps, under the command of German Field Marshal Erwin Rommel, the crafty "Desert Fox," offered stiff opposition to Allied forces.

Now, intelligence activities would center on terrain studies and German and Italian orders of battle in Tunisia. There was a lot of work to be done.

"Rommel was a brilliant commander," the general said. "We knew his armored divisions were superbly trained. If we were going to face Rommel's panzers, we needed every advantage we could get."

"Probably the last German general you'd like to face, right? Rommel had a tough reputation."

"He was one of the best, and he was a master of tank warfare."

I said, "Candidly, how did it make you feel?"

"I'd say it just made us more determined to do our job. We knew very well that when you went up against Rommel there was not a lot of room for miscalculation."

Results of the intelligence staff's work was needed even sooner than expected. When the American II Corps was defeated at the Kasserine Pass, Eisenhower ordered Patton to take over that unit. The II Corps, in turn, was placed under direct command of British General Sir Harold R. L. G. Alexander's 18th Army Group.

Patton called on Oscar Koch to be his intelligence officer, but accepted Koch's recommendation that Colonel Benjamin A. ("Monk") Dickson, the existing Corps G–2, be retained as chief of staff for intelligence. Colonel Koch would be assistant G–2.

Facing Rommel's Fifth Panzer Army with the luxury of shared intelligence responsibilities, Koch was able to concentrate on the techniques of Rommel's tanks on the ground, in combat. He was immensely impressed, but not surprised, by what he found.

"Lure and entrapment by double envelopment was a favorite technique," he wrote later. Rommel obviously knew that about two–thirds of the American force usually would be making an assault, with the remaining third in reserve, and planned to meet an attack accordingly.

"We learned that Rommel's tanks would move slowly to avoid dust and noise—so slowly that unless checked carefully they might be judged immobile. In battle they might stop and appear to be hit, only to reopen with rapid and accurate fire once the attacker had shifted to another target."

Col. Oscar Koch, left, and Lt. Col. Claude Burch, U.S. Seventh Army provost marshal, at Segesta, Sicily, in 1943.

Col. Oscar Koch in undated U.S. Army photo.

ANNEX 7

PRE D-DAY PLAN OF PHOTO RECONNAISSANCE

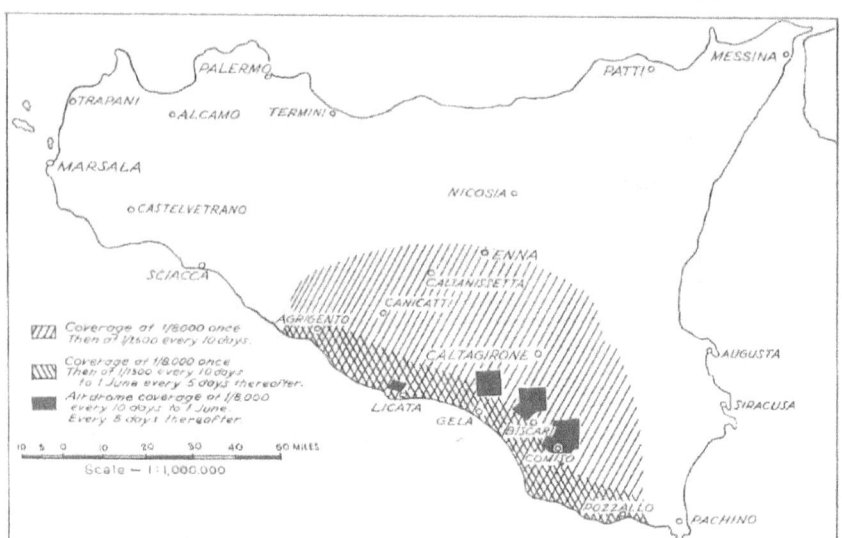

Col. Oscar Koch, as U.S. Seventh Army G–2, was the first on Patton's staff to begin planning for Operation Husky, the invasion of Sicily. Because intelligence was seen as the foundation for all other preparation, Col. Koch put heavy emphasis on air reconnaissance, as reflected in this map from the Sicily after–battle report.

Hundreds of thousands of propaganda leaflets aimed at persuading both German and Italian soldiers to surrender were distributed by the U.S. Seventh Army G–2 Section during action in Sicily.

Engagement with Rommel's forces was close at hand. With the Allies, including Patton's II Corps, moving toward it and nowhere to run, the German 10th Panzer Division would fight desperately to head off further Allied advance. The II Corps met Rommel's tanks head-on at the Battle of El Guettar. They'd been forewarned by intelligence, and called in artillery and bombing that quickly took a heavy toll on the German armor. The Germans regrouped and launched a second attack late in the afternoon, but it, too, was turned back.

Although Rommel actually was back in Europe on sick leave at the time, his subordinates were well versed in his strategy of tank warfare. What the II Corps G–2 Section had learned about his tactics was put to good use. The Battle of El Guettar was an important victory for the Allies. For the forces of Hitler and Mussolini, it was a precursor of what lay ahead for them as long as fighting in North Africa continued.

In Tunisia, psychological warfare still was a function of the G–2 Section. The intelligence staff had the advantage of knowing that German Field Marshal Friedrich Von Paulus had surrendered at Stalingrad and the Russians had taken 91,000 German prisoners in a single ten–day period. The Germans also had lost 750 aircraft and thousands of tanks, artillery pieces, and smaller weapons.[4]

The psychological warfare team assumed this kind of negative information had been withheld from the German troops in Africa. Certainly it would have a demoralizing effect if they knew.

"We dropped leaflets that urged the German troops to surrender," the general explained. "We told them their cause was lost. We emphasized the German losses on the

Russian front in particular, guessing they hadn't been told about any of this. We let them know that their forces advancing on Stalingrad had been soundly defeated."

As to the current front, "We showered the Afrika Korps with leaflets emphasizing the tremendous amount of North African territory they'd already lost to the British. Fresh Allied troops were advancing on them from the west. Germany was under constant bombing attacks. Their families back home were dying in these raids, which would only get worse as the war dragged on. They could help end it all by surrendering now."

"Pretty tough stuff for soldiers fighting a long way from home," I said.

"Of course it was. We wanted them to feel like they were fighting in vain, risking their lives in a war they couldn't win. And knowing that families back home are being put at greater risk every day the war goes on would be really rough on any soldier."

"And the Italians?"

"We told the Italians that Hitler and Mussolini were grasping at straws and would take the Italian troops down with them. We knew the Italians hadn't been treated too well by the Germans, and considering the fact that Mussolini had tied in with Hitler we wanted to make them think about giving their lives for Hitler and realize they had nothing to gain by continuing the fight. And they didn't have to. All they had to do was surrender."

Leaflets dropped on the enemy almost always offered safe passage to any soldier who wanted to surrender. The German and Italian troops were promised that if they surrendered they would be well treated and well fed. The propaganda materials painted a picture of life as a POW being much better than life as a soldier in combat fighting for a losing cause.

It soon became clear that these propaganda tactics were effective. The success of this activity in Tunisia led to greatly increased propaganda efforts later, during Patton's U.S. Seventh Army campaign in Sicily.

General Koch fully sympathized with Patton's attitude toward prisoners of war. Patton had made clear years earlier, when he was executive officer of the Cavalry School at Fort Riley, that prisoners should be treated humanely. When an instructor referred to PW cages, the commonly used term, Patton said "cages" connoted captured animals. Prisoners of war were men. Hereafter, the reference would be to PW "enclosures."

"The Old Man always claimed that getting captured wasn't the soldier's fault, but the commander's," the general said. "Command had failed, not the man taken prisoner. And he believed this, which I think was one reason he was a good commander himself."

Regardless of where the fault lay, the G–2 Section looked on captured enemy soldiers as a vital source of intelligence. Information gained from prisoners sometimes proved useless, but it often turned out to be of great importance.

The general said knowledge gained from POW interrogations was especially useful in the weeks preceding the Battle of the Bulge. One prisoner, a high-ranking German officer, revealed to interrogators he'd been told three months earlier that a large counterattack was scheduled for the end of December. Another claimed Hitler recently had called on the German troops to fight hard to hold their positions because in two weeks a counterattack would be launched.

Information as simple as a captured soldier's unit could be very helpful. In one instance, locating the enemy's Sixth Panzer Army was vital in estimating how large the German buildup was in the weeks before the Bulge.

"Even when we already had the information through other intelligence sources, as we often did, confirmation from a POW gave us just that much more confidence in what we had," he explained. "You can never reject one more source."

I earnestly looked forward to plunging into General Koch's intelligence activities related to the Battle of the Bulge and would have jumped ahead gladly. He declined. "All in good time," he said. "We don't want to mess up our clever scheme of organization before we even get our teeth into it."

I knew he was right.

As to POW stories, my favorite one, told in detail in *G-2: Intelligence for Patton*, concerned an enemy agent captured in France. He was a young Frenchman who had switched his allegiance to the Germans after they had occupied his homeland for some period of time.[5] He had been dropped into the Metz area by parachute on a moonless night, packing a heavy radio, and was lost from the moment he hit the ground. He soon was captured and needed little encouragement to tell his story.

He felt no strong loyalty to the Germans and was ready to quit. Would he be willing to help the Allies? Yes. Just tell him what to do.

He was told to report back to his German contact and say he'd landed safely. Ask for further instructions. Then, he would tell his captors immediately whatever his superiors ordered him to do next. Koch and his staff took it from there. They gave him correct but harmless information to relay to the Germans in response to his initial orders.

When his contact directed him to meet a second agent, a team of American soldiers went in his place. The young Frenchman continued to report faithfully to his home base, and finally was told to meet a number of other agents who would arrive at certain times and places.

"Except for one poor stray, they all landed on schedule," the general said, "and we made sure to have American reception committees there to meet every one of them. It took us two days to find the stray."

Then General Koch burst out laughing. "Here's the part you'll really like," he said. "Before we'd finished with all this, this guy is told he's been awarded a medal—the Iron Cross. His superiors said it was for devotion to duty and the Fatherland."

The general made no bones about his attitude toward spies overall. "I would never trust them," he said flatly. "If they change allegiance once, who's to say they won't do it again. Any information we got from a spy was considered useless until we could check it out. I don't mean we ignored it. We always took it seriously and checked it out as soon as we could. Sometimes it turned out to be very useful, and sometimes we just got wild stories."

More and more, I was coming to see why Oscar Koch had been outstanding in his role as head of Patton's intelligence section. His devotion to detail, his emphasis on using all sources of information available, his insistence on accuracy, his lack of personal ego involvement—all these were traits I had come to recognize in the general in even the relatively short time I'd been working with him.

And there was another quality, surely highly useful in

his years in combat intelligence, which was extremely helpful to me. He had excellent writing skills and summarized facts succinctly, much as I had learned to do as a journalist. There was never a single instance in which I needed to ask for clarification of something he'd written. The general often drafted material for the book in progress, and although he usually referred to it as a "rough" draft, I always found it well done. In summing up the situation at the end of the Tunisian campaign, he "drafted" the following paragraphs. They appear in *G–2: Intelligence for Patton*, precisely as he wrote them:

> Several things were by this time becoming self-evident about our intelligence operations. If there had been a feeling that intelligence in combat was a matter of crystal-gazing, or that some occult power could be called upon to give the answers, those ideas were dispelled. There was no longer any mystery about it; intelligence took conscientious application and hard work.
>
> The basic principles of the few manuals and other publications on combat intelligence then available were sound: collection, collation, evaluation, interpretation, and dissemination. Practice had shown that initiative and imagination on the part of intelligence personnel were necessary for successful application of the principles taught.[6]

• This had a familiar ring. This was military precision. This was "by the numbers."

As I had learned in my own combat infantry training, this was effective teaching. Here was the simple formula for intelligence in combat. Anyone could do it, right?

Well, perhaps not. The inexperienced intelligence officer could not be expected to have the skill with which Oscar Koch applied the principles; this was the "initiative and imagination" he called for. As I worked with the general over the weeks and months to come I never would cease to be captivated by his record of success in this most demanding and unforgiving arena. Combat intelligence might be a craft, but Oscar Koch was an artist.

I was pleased with the progress we were making. Our book was beginning to take form. There still was much work to be done, work that became more important to me with every passing week. I found immense satisfaction in watching the general's growing confidence that our efforts would lead to completion and, with good fortune, publication of a book that would be the culmination of his long-time hopes and let him finally reach the elusive goal he had set years before.

Neither of us would have imagined at this point the dreadful impediment that lay ahead.

Chapter 7

ON ONE OF MY frequent visits to the Koch home, the general called attention to something we had in common. Something important. We both served in the U.S. Third Army. The general still had Third Army in his blood and rightfully so, because he made an immense contribution to its glorious record under Patton during World War II. It was Third Army that swept across France, saved the Allied cause in the Battle of the Bulge, and accepted the surrender of thousands of enemy troops as it raced toward Berlin. And Oscar Koch's brilliant work as Third Army G–2 was among the foremost reasons for its success.

Third Army was no longer an active combat unit at the time I was stationed at Fort Jackson, and I attached little if any significance to the fact I was on a Third Army post. Later, as I came to know Oscar Koch, I was happy I'd worn that distinctive red, white, and blue circular Third Army patch on my sleeve and happy he still considered this noteworthy.

But even though I was ignorant of Third Army history at the time, Fort Jackson, South Carolina, adjacent to the city of Columbia, was a destination exotic enough for a Midwesterner who never had been that far from home. I often had been enthralled by my great-aunt Emma's stories of how her people, those of my grandmother White, had made their way to southern Illinois from North Carolina. I had romanticized her stories and felt as if I had a touch of Southern blood. Orders to Fort Jackson were like a ticket to the search for family roots.

It was spring and the first time I saw Columbia, the state capital and home of the University of South Carolina, the azaleas were in full bloom and Columbia was one of the most beautiful cities I'd ever hope to see. It still is, but like so many others it's too commercialized today. Many of the quiet, tree-lined streets of the 1950s are four-lane highways now, lined instead with fast-food restaurants. The area is drowning in traffic.

My three friends and I traveled from Fort Leonard Wood to Columbia by train, on military travel orders. The trip began with a painfully slow troop train to St. Louis, where outbound passenger trains heading in all directions were waiting. Some, we were told, had been held as much as an hour by our late arrival. But they waited, because the Army surely gave them a great deal of business and they wanted to be accommodating.

After St. Louis, we rode first class. Our train slipped smoothly eastward over the L&N Railroad's tracks across southern Illinois. It made a brief stop in my home town, Carmi, and I felt a tinge of homesickness as I sat and looked out the window, watching a couple of passengers get off the train and another get on, and wondered how long it would be before I was here again.

We had a compartment overnight to Atlanta, then a

change of trains to Columbia by way of Greenville–Spartanburg. This struck us as the long way around, but we thoroughly enjoyed the trip. Our car was comfortably filled with congenial travelers and the striking little monadnocks that began to pop up at random and tell us we were headed into the Appalachians made it a pleasure simply to ride along and watch the changing scenery of northeast Georgia through the wide railcar windows.

We got there a day early. A night at the YMCA sounded better than an extra night at Fort Jackson, especially since we weren't expected yet. All the beds at the YMCA were filled, though, and we stayed at a cheerful Columbia boarding house. Late the next day a taxi took us to the post.

Fort Jackson had been home to the 101st Airborne Division until just before I got there in the spring of 1956. With the 101st relocated to Fort Campbell, Kentucky, Fort Jackson had become the Third Army's infantry training center. The famous Screaming Eagle insignia of the 101st, more popularly the "puking buzzard" to non–airborne, still marked almost everything on the post, and even the local taxi drivers weren't familiar yet with the new unit designations.

Third Training Regiment? It took my driver a while to find out where it was.

Insofar as buildings were concerned, Fort Jackson looked just like Fort Leonard Wood. The low, flat one–story office structures and the two–story barracks all were clapboard–sided and painted a yellowish tan. The similarities ended here, though, as my new post was heavily forested with tall pines and landscaped with shrubbery of varieties that would have been foreign to the Missouri Ozarks. I saw the beautiful crape myrtle for the first time—though I have it now in my yard in southern Illinois—and I loved the ubiquitous mimosa.

And there was something else I'd never seen before. Kudzu vines. Everywhere, overgrowing everything in their path. The Army apparently hadn't developed a plan of battle this forceful invasive species and I don't know if it ever has. I was familiar with the vigorously spreading vine honeysuckle in southern Illinois, but compared with Kudzu it would be an impotent imposter in its tendency to try and suffocate everything that grew nearby.

I was fortunate again, arriving just as the Third's regimental special orders clerk was about to be discharged. Instead of getting sent down to one of the battalions to become a company clerk—think Corporal Radar O'Reilly in the television series, "Mash"—I was kept at regimental headquarters to take over what surely must have been one of the most cushiony jobs in the entire U.S. military establishment.

The decision to keep me was made by Mr. Quist, a warrant officer who was assistant regimental executive officer. I wanted to think that he had latched onto me because of my obvious potential as a first-rate clerk-typist, but Sergeant Mason Sykes, who was the top administrative clerk and who would become a good friend, said it was because I impressed Mr. Quist with my military bearing.

Military bearing? This was the first and probably the only time I ever was accused of even having military bearing, much less impressing anyone with it. The truth is, I had blundered through again. I'd never seen warrant officer brass before and didn't realize Mr. Quist was an officer when I first reported in at his desk. I realized my error after I'd handed him my papers and tried to pull myself into a position of attention without being too obvious.

Apparently I was just obvious enough. Mr. Quist noticed. And he was impressed! That was military bearing.

Mr. Quist pretended to be a stickler for rules and regulations, but actually was soft at heart. He would be the one who signed off on my work, and although at times he could be somewhat over-demanding, he'd always listen to reason and on the whole he was easy to work for.

My new home, the Third Training Regiment, was Fort Jackson's advanced infantry training unit. The regimental commander was Colonel Donald Blackburn.

Although we were in the same building, I saw very little of him. But I quickly learned of his reputation from World War II, when he was a young lieutenant left behind when General Douglas MacArthur abandoned the Philippines. His diary of daily activities organizing and commanding a force of native guerrilla fighters to harass the Japanese troops was the basis for the 1955 book, *Blackburn's Headhunters*, by Philip Harkins.[1] The book, in turn, was made into a B-grade movie, "Surrender, Hell," in 1959. Keith Andes played the role of Blackburn while the colonel himself served as a technical advisor on the film. I have neither read the book nor seen the movie.

I tried several times to get the book from the Fort Jackson post library, but it was never available. Popular reading among junior officers and NCOs of the Third Training Regiment, I suppose.

Colonel Blackburn's experience in the Philippines left him well-qualified for more important assignments, particularly in the field of clandestine warfare, and he regained some of his World War II luster in Vietnam. He was promoted to brigadier general and played significant command and development roles in the Army's Special Forces.[2]

I had other regimental commanders over time, but Colonel Blackburn is the only one I recall. It seemed that commanding officers found little need to check in with me and I did my best to remain anonymous to them.

Two memorable things happened not long after I arrived at Fort Jackson, one of them life-altering and the other merely juicy fodder for "old Army days" stories. The life-altering event was when I met Mary Corley, a beautiful University of South Carolina student from Columbia, and the lesser event was the reappearance of General Normando A. Costello.

Mary Corley didn't date soldiers, but—well, she did. It was a double date with one of her friends who was dating one of my Army buddies. We enjoyed each other's company. She soon introduced me to rock and roll music, of which she had an enviable 45-rpm record collection—Little Richard, Fats Domino, Chuck Berry, Ray Charles, all the greats of the time. Our tastes were broad, though, and we soon discovered that we shared a love of Mario Lanza. We took this to be a clear omen that we belonged together, developed a quick bond of devotion, and were married some eighteen months later.

As to General Costello, the much feared assistant commander at Fort Leonard Wood was assigned to Fort Jackson as the new post commander. He put his stamp on things pretty quickly. Even office boys like me were required to change out of our comfortable khakis and wear the field uniform, including combat boots and helmet liner. Further, we had to get epaulets sewn onto all our fatigue shirts at our own expense. I just happened to be going home on leave for two weeks and my mother sewed epaulets on my shirts while I was there.

Mother didn't charge me for her sewing.

There were a few other new wrinkles. Stories quickly spread about the commander dropping by the post exchanges during working hours and catching enlisted men sloughing off, taking coffee breaks or whatever, and putting them on immediate orders to Korea. The same fate awaited any soldier walking along a street on the base who failed to salute when he met the general's automobile. Some of the things we heard probably were exaggerated, but I and a few other Third Training Regiment clerks promptly abandoned our morning coffee breaks at the main post exchange just in case.

None of us knew, or would have cared very much, that Costello was a West Point graduate and had been awarded a Silver Star for gallantry in World War II. A colonel, he was commander of the 319th Infantry Regiment, 80th Infantry Division. He went on to receive a Distinguished Service Medal for actions during the Vietnam War and retired as a major general.[3]

Long after he served as Fort Jackson commander, Costello's daughter, Toni, recalled an incident that took place there late in the afternoon on the day before her father took over. She was 12 years old at the time, riding in the back seat of the family automobile when he drove up to the main gate at his new post, dressed in civilian clothes, and asked the young MP on duty if something special was going on. The soldier said yes, they were getting a new commanding general in the morning.

Did he know the new man's name? Costello asked.

The MP said it was General Normando Costello.

The general replied that he knew this name, and added a word of caution. Everyone had better be prepared, he warned, "because I've heard he's the meanest son–of–a–gun in the Army!"4

There obviously was some truth in his statement, at least the part about us being prepared. We weren't. There would be changes we were not ready for.

When I read this story today, it occurs to me that Patton would not have said "son–of–a–gun." His use of colorful language was a Patton trademark. But General Koch, never a Patton detractor, told a group of Air Force Reserve Officer Training Corps cadets at Southern Illinois University his old commander was "profane but never vulgar."

Patton's propensity for using strong language apparently never rubbed off on Oscar Koch. The general's cousin, Dr. Murray Zimmerman, claimed that Koch once referred to Colonel Otto Skorzeny, the German commando leader, as "that son–of–a–bitch." But the most off–color remark I ever heard him make was, "Look at this poor little bustard. He doesn't look like he found much glory in fighting for Hitler."

He spoke in a decidedly sympathetic tone, shaking his head somberly, addressing a photograph of a bedraggled and scared–looking teenaged German soldier marching in a long line of enemy troops who had just been captured by Patton's forces a few weeks before the war ended.

I particularly like this Koch quotation because it demonstrates two qualities I came to take for granted: He would not use an even moderately vulgar word, and he was deeply concerned over the victims of war. When we got into his Third Army intelligence activities preceding the Battle of the Bulge, I found he held no personal animosity toward

Bradley and Eisenhower for ignorjng his reports. But he was bitter because of the resulting casualties, many of which he felt could have been avoided.

Koch's Third Army war room always featured a daily casualty chart. The chart showed both the Third Army and enemy losses, and Patton studied the casualty chart closely every day. It was the first thing he looked at in the daily war room briefing. Third Army was the first headquarters to post these numbers.[5]

General Koch was proud of the Third Army war room. He showed me pictures, claimed that every facet of it was entirely functional. It struck me that this no doubt was an example of how he went the extra mile to do the best he could in all areas of responsibility. The Third Army war room gained recognition as a superior example among the military commands in Europe.

We seldom got into old Army stories, but the general and I decided we did have at least one common military experience in addition to our Third Army service: our slowness in making rank.

This was a joke, of course, because my experience concerned promotion to private first-class and his related to being made brigadier general. The Department of Defense had a one-year freeze on promotions at the time I was drafted, and General Koch had a hard time gaining his first star because he was a staff officer not in command of troops.

We also had in common the support of our superiors. On the very day I became eligible for higher rank the regimental first sergeant, a tall, mild-mannered Georgia career soldier named Jesse Cross, walked over to my desk and laid a sheet of paper in front of me. It was a court martial report on a private first-class down in one of the companies.

"Take his stripe away and give it to yourself, Hays," he said.

I wrote my own promotion orders.

Ninety days later, the first day regulations allowed me to be advanced again, Sergeant Cross directed me to promote myself to specialist third-class. It was fortunate he was keeping track of the dates. I certainly was not. Rank was not important, although each promotion added a few dollars a month to my modest Army pay.

Needless to say, General Koch's situation was a bit more complicated. Patton's principal staff officers who held the rank of lieutenant colonel soon became colonels and some of them eventually gained their stars. Others, including Koch, did not. Patton repeatedly recommended them for promotion to brigadier but his recommendations were repeatedly ignored.

Finally, at a time when his two European superiors, Bradley and Eisenhower, were away temporarily, Patton sent a letter to the War Department in Washington once again calling for his staff colonels to be upgraded to general rank. He signed the letter as Third Army commander and then added favorable endorsements as acting Army Group commander (for Bradley) and acting Supreme Allied Commander (for Eisenhower). He even added a personal, handwritten note to the Secretary of War, whom he knew well. Once again, he got no results. The war was over before Koch was promoted to brigadier general.

I never sensed any resentment over his slowness to gain general rank on Koch's part. His main concern was that his advancement might have been held back because of his staff position, G–2. Once again, he feared, it was a simple case of intelligence not getting the respect it deserved. Not everyone in a position to make such decisions gave intelligence the high priority that Patton did.

Chapter 8

IN MID–MARCH 1966, not long before I first met General Oscar Koch, brutal race riots erupted in the Watts section of Los Angeles. Two weeks later, an estimated twenty-five thousand antiwar demonstrators marched in the streets of New York City. The next several months brought serious race riots in Chicago, Cleveland, Omaha, and smaller cities across the country. The National Guard was called out to help quell riots in Omaha, and Ohio Governor James Rhodes declared a state of emergency in Cleveland. The Reverend Martin Luther King Jr. was pelted with rocks while taking part in a civil rights march in Chicago.

Such events seriously impacted the national psyche, and the recognition that we lived in such troubled times weighed on the general and me as we worked on *G–2.* in the months to come We tried to push all this aside and not let it affect our efforts, but sometimes it played a role in where we focused our attention.

Looking over statistics from Patton's operations in Sicily, I saw that some of the after–battle report data included numbers separated into columns with racial designations. Medical Service reports on hospital admissions and diseases contracted by the soldiers, for example, were shown in separate charts under the headings "white" and "colored."

"This surprises me," I told the general.

He looked at the site I indicated. "You have to remember, this was 1943," he said. "This was still the Army's way. We still had a long way to go on this kind of thing."

I considered the date, and realized that more than a decade later, during my term of military service, the Army still classified individuals by race. Every day when I wrote special orders covering troops in or coming to the Third Training Regiment at Fort Jackson, each soldier was identified by name, rank, serial number, and race, "CAU" for Caucasian and "NEG" for Negro, even though the military services had been fully integrated for some time.

Patton was accused of racism, although General Koch thought this charge was unjustified. His view was that Patton, like others his age, had grown up in a segregated society and had known and come to accept some of the common black stereotypes. He pointed out that his old commander integrated rifle companies, the prime example of close association during combat. So far as he could see, Patton "had no problem putting white and black troops together."

Patton asked that the 761st Tank Battalion, a Negro unit, be assigned to his command late in the war. According to a history of that unit, Patton told the men, "I would never have asked for you if you weren't good."[1] Although Patton wrote that "a colored soldier cannot think fast enough to fight in armor," Hirshson claimed that the Third Army commander "found much to admire in his black units."[2]

It is hard for us to remember today the extent to which racial segregation existed in the era of World War II. After the war, in 1948—the same year President Harry S. Truman ordered the full integration of black and white troops in the military—Pulitzer Prize-winning newspaper reporter Ray Sprigle drew national attention with his story about Macy Yost Snipes, a young black veteran in Georgia who was murdered for being the first Negro in his town to vote.

"Death missed him on a dozen bloody battlefields overseas, where he served his country well," Sprigle wrote. "He came home to die in the littered door-yard of his boyhood home because he thought that freedom was for all Americans, and he tried to prove it."[3]

General Koch said change in the racial makeup of Army units was "fairly obvious" even in the relatively short span of time between World War II and the Korean War. I would have thought so, because when I was drafted not long after the latter conflict ended all my units were made up of both white and black troops.

Change within the military is based on orders that come from the top down, as was the case with Truman's command to fully integrate black and white troops. This means that change is likely to come much faster in the military than it is in civilian America. This is important, because with troops in combat or preparing for combat there's little room for gradual transformation.

American troops were in combat almost continuously from World War II through the end of combat in Vietnam. While at Fort Jackson in 1956, I served with NCOs who had been called to duty near the end of World War II, then recalled for combat in Korea. Their civilian lives had been thoroughly disrupted by the time that conflict was over, they had put in several years of military duty, and regardless of their original intentions they had decided to make the Army a career. They were dedicated professional soldiers, good at what they did.

Simmering unrest in the Mideast was serious enough that at one point during my tour of active Army duty, all military personnel due for release were "extended" for an extra six months. President Eisenhower would later send a strong force of U.S. Marines into Lebanon. And by the mid-1950s America had commenced its long entanglement in Vietnam. With war or the threat of war almost a constant, there never was a time during this entire period when the American armed services would have assumed the luxury of a gradual phase-in of racial integration.

Military service inevitably brings change, and learning to adapt should be viewed as a strength in an individual soldier. The Mexican border region must have seemed terribly exotic to young Oscar Koch from Milwaukee. My experience was hardly comparable, yet Fort Jackson's setting in the Deep South led me to experience a way of life that was foreign to anything I'd seen before.

Many of the training cadre there—the drill sergeants and instructors—were paratroops who'd served in Korea, holdovers from the 101st Airborne Division. They included black soldiers who faced a totally segregated society the instant they stepped off the base.

If they were native Southerners themselves, this obviously wasn't something new. But a substantial number of them were from the Northeast, cities like Boston and New York and Philadelphia, and until they were stationed in the South they may never have experienced the demoralizing and dehumanizing circumstances of official, government-sanctioned separation of the races.

I never actually had occasion to witness how my African-American friends on the post were treated by white society in Columbia and elsewhere around the base because there were few places I and they went together. My only direct experience with segregation came when the University of South Carolina refused to accept me and other Fort Jackson soldiers into classes the Army would pay for because it would not accept black students and the Army said it had to take everyone.

Fortunately, Columbia College, a small but excellent women's liberal arts college, stepped in to fill the gap. I was able to take two classes there, and thought they both were very good.

I got to Fort Jackson nearly two years after the U.S. Supreme Court's Brown v. Board of Education decision outlawed racial segregation in the public schools, but little progress toward school integration was apparent in the Deep South. The biggest impact of the court's verdict, as far as I could tell, was the great number of "Impeach Earl Warren" billboards along the major Southern highways. Warren, of course, was Chief Justice of the U.S. when the court delivered the ruling.

And I arrived in the South only a few months after a jury in Mississippi had acquitted a pair of white men charged with the murder of Emmett Till, the black youth from Chicago they believed had flirted with a white woman. The two men later confessed their guilt and described their activities in gruesome detail to a magazine writer who managed to get them paid for their story.

My own most vivid memory relating to race relations in the South at the time stem from an experience I had on a sultry summer evening in 1956 when, for the first time, I fully comprehended the awful fear of authorities that must have been an everyday factor in the lives of black citizens. I was a 21-year-old soldier hitchhiking across the tranquil Piedmont region of South Carolina, on my way home from Fort Jackson to southern Illinois.

In 1956, almost any driver would stop for a man in uniform. A young mother with two small children in the back seat gave me a ride in eastern Kentucky and apologized that she couldn't take me farther. A farmer in Tennessee, somewhere between Knoxville and Nashville, went far out of his way rather than leave me stranded on a rainy night. Many drivers offered to buy me meals or asked if I needed money.

This was the American South in the 1950s. The good South. There was a darker side.

On this evening I recall so clearly, I was standing alongside U.S. 276 somewhere north of Greenville when a car slowed and went past, then stopped and backed up. A man on the passenger side rolled his window down and told me, somewhat timidly it seemed, that I was welcome to ride "if you want to." He was a middle-aged black man—I don't think many used the term "African-American" then—and the driver also was a black man, a bit younger.

I said I'd welcome the ride. He slid over and I crawled in beside him. One of the men asked where I was from and when I said Illinois they seemed relieved. One of them said, "Then you know what it's all about." I knew what he meant and I hadn't the heart to tell them I had seen my share of racism in my home state. But the racism I had experienced in Illinois was different. We didn't have segregation as an official policy, sanctioned by law and enforced by authority. And in any case I, as a white man, would never experience the crushing weight of such discrimination. Those two men probably had.

They were a congenial pair and I enjoyed our conversation. Then I sensed the driver growing tense. He leaned forward, his arms rigid, gripping the steering wheel as if afraid he might lose control. We had just entered the town limits of Traveler's Rest, a small burg close to the North Carolina state line. There was little traffic. The speed limit was clearly posted and the driver made a point of the fact that he was well within it.

"There's a policeman right behind us," he said, focusing on his rear–view mirror. "I can see his police cap in the lights coming up behind him."

The man next to me virtually froze in his seat. "Don't go too slow and look suspicious," he told the driver. "Try to stay right on the limit. What's he doing now?"

"He's just sticking on our bumper," the driver said. "He's going to stop us for sure."

The two black men's fear was palpable and overwhelming, unlike anything I'd ever witnessed before. It was obvious they were terrified at the prospect of being stopped by police. The driver grew increasingly nervous, afraid either to speed up or to slow down, and the man next to me began to shake.

We were half way through Traveler's Rest when the car behind us pulled out and went around. As it drove past, we all could see that the driver wore a then–common gas station attendant's cap. Texaco, or whatever. It would be quite easy to mistake for a policeman's cap seen in profile in a rear–view mirror. My two companions' relief was as apparent as their fear had been only minutes before.

Let me admit that I know nothing about these two men.

They may have been wanted criminals. Bank robbers, for all I know. Perhaps serial killers. But I don't think so. I think they were ordinary African-American citizens of South Carolina who understood that white police officers were not their friends, perhaps having learned from their own personal experiences or maybe simply having heard the stories of others.

We know now that, a few years later, black members of the "Freedom Riders" who journeyed to the South to call attention to illegal racial segregation in public interstate transportation facilities such as bus stations were subjected to violence through the treachery of white policemen. I suspect this news did not come as a great surprise to the two men who graciously stopped and offered a ride to a soldier in uniform that night on a busy Southern highway.

General Oscar Koch spent the greater part of his life defending freedom, and had strong feelings about the rights of citizens. Further, he pointed out that segregation in the armed forces, even if it had been fair, came at a high price. It led to a waste of manpower, for one thing, and also a costly duplication of training facilities.

A ready example is The Tuskegee Airmen. They carried this name simply because a separate training base was established for them at Tuskegee, Alabama.

I was reminded of this recently when the movie, "Red Tails," opened in theatres around the nation. It was about the Tuskegee Airmen and generated new discussion of the role racial segregation played in their selection and training.

I hadn't known that the 99th Pursuit Squadron, the first unit of the Tuskegee Airmen, was activated not in Alabama but rather at Chanute Field at Rantoul, Illinois.

The men of the 99th had one of the highest cumulative grade–point averages of any group stationed at Chanute.[4] They eventually were transferred to Tuskegee for training, after the installation there was established as a separate, segregated base for black servicemen.

It was a full decade after my experience as a hitchhiker in the South that I met and began work with General Koch. The armed forces, under Truman's directive, had long since been fully integrated. The American civilian society had not. The year 1966 ended just as it had begun, with ugly conflict between black and white citizens.

In other facets of society, life went on as usual.

Medicare went into effect. The National Organization for Women was founded. Dr. Seuss's "How the Grinch Stole Christmas" was aired for the first time on network television. Singer Neil Diamond's recording of "Cherry, Cherry" landed on the popularity charts, Janis Joplin played her first West Coast performance, and Jimi Hendrix wrote "Purple Haze." A movie actor named Ronald Reagan was elected governor of California, and the U.S. Supreme Court declined to block relocation of the major league baseball Braves from Oscar Koch's home town of Milwaukee to a new home in Atlanta.

While we were not blind to the world around us, General Koch and I were careful not let any of these happenings become serious distractions. We had much work to do.

Chapter 9

THE ANCIENT GREEK lyric poet Pindar wrote that, "Unsung, the noblest deeds will die." I do not want Oscar Koch's noble deeds to be lost to history. His contributions are far too many and much too important for him not to receive the credit he deserves.

I never believed the general wanted to be recognized as a hero. He did not pretend to be important beyond any other veteran of the armed forces who had served to the best of his or her ability, regardless of duties performed. He wanted, primarily, for his work to be acknowledged as a means of stressing the importance of intelligence in combat. He had seen too many instances of what he did being misstated, or not mentioned at all, by those writing the history of World War II. If what he did was not important, intelligence was not important. He knew better.

Secondarily, I think he wanted his story to be told so as to bring more credit to Patton. He felt strongly that everything he did was done as part of a team. Patton's team. Patton's G–2 Section. He would be the first to emphasize that the intelligence reports themselves meant little until they were acted on by the commander. The same information that Patton had in the weeks preceding the Battle of the Bulge was available to Eisenhower and Bradley, but they did not act on it. Fortunately, Patton did, otherwise the outcome might have been far worse.

Despite his remarkable accomplishments as G–2 for one of the most celebrated generals in American history, Oscar Koch held no lofty opinion of himself. He preferred to give the credit to the infantrymen on the ground, the paratroops, Navy gunners, the pilots of the bombers and fighter planes who risked their lives in support of a mission set by higher authority.

"They faced enemy fire," the general told me one day. "And I didn't." Simple as that, although I think this significantly understates his own role in battle during three wars.

And just as he did not seek fame, it always seemed to me that Oscar Koch did not expect much in the way of material things. Not surprising, for a military career typically doesn't lead to a substantial accumulation of personal items to be moved from post to post and Army salaries were not high. He retired as a brigadier general, yes, but he had spent most of his career at relatively low rank and he was contented to take what the Army offered. As he once put it himself, he "never joined the Army to get rich."

Dr. Murray Zimmerman, Koch's cousin, told of an incident in London after the war that precisely matched my view of the general's lifestyle and personality.

"I remember going to Locke in London and buying a homburg [hat], which cost about twenty dollars then," Dr. Zimmerman said. "Oscar admired it and tried it on. It fit him perfectly and he looked very distinguished in it, so I went back to Locke and bought one for him, too. He was absolutely ecstatic with joy. He was as delighted as a kid with a new toy. I don't think many people in life had done any altruistic things for him, though he deserved greatly to be treated kindly."[1]

When the general told stories about his life in the Army, they most likely were to be common, everyday-soldier types of anecdotes. What happened to me as a draftee regimental special orders clerk who never came close to combat rated as highly with him as his experiences on the front lines of the war in Europe under Patton. It wasn't that I had important things to tell; rather, it was because they were my personal stories and he valued them accordingly.

General Koch's wonderful sense of humor bubbled to the surface easily. He was always witty, and he appreciated wittiness in others. This was a trait he shared with John W. Allen and I've often thought what marvelous discussions surely went on between them. I came nowhere near their standard on wittiness, but it was fun to try.

The general and I both were tired one evening after a longer than usual work session and began to attach deliberately exaggerated importance to anything that offered a break from the nuts and bolts of combat intelligence. He had spent the last several minutes explaining to me the differences between the British intelligence model and the American, the most significant of which was the intelligence officer's end goal. The British officer was expected to predict the enemy's *intentions* for his commander, while the American officer's responsibility was to present his commander with the enemy's *capabilities* and let the commander decide on intentions.

I had been unusually slow to pick up on some of the finer points of the general's examples. I finally caught up and assured him I understood the significance of this distinction. But I told him the whole discussion reminded me of the ancient story of the failure of Croesus of Lydia to comprehend the prophecy of the Oracle of Delphi.

"And you're going to tell me the story," he said with feigned resignation, as if he had no interest in the subject but accepted the fact that he was going to have to listen.

"Of course." I suspected he knew the story as well as I did, but I would play out my game.

Croesus, I related, is said to have favored the Oracle with a fortune in gifts and then asked whether he should invade a neighboring country. The Oracle said if he went to war he would "cause the destruction of a great empire."

Croesus invaded the neighboring lands and not only was defeated but was taken prisoner. He sent word to the Oracle complaining that he'd been misled. The Oracle's response was that he had not been misled; he'd been told there would be the destruction of a great empire and there was. His own.

"So," I concluded, "Croesus missed the boat on his enemy's capability, right?"

"You know what his problem was, don't you?"

"No," I said, "except maybe—"

"He didn't have an intelligence officer. Oracles pick things out of thin air. Only the big things. Intelligence officers have to work harder and give their commanders the whole picture, including the small details."

The general almost never interrupted when someone else was speaking. It made me feel good that he not only took up my little game, but actually got enthusiastic enough to break in when I was talking. And even if I had led us into a somewhat ridiculous discussion of ancient oracle myths he did not and would not make me feel foolish.

Our work sessions usually were at the Koch home because that's where his files and other materials were. He made it clear he was willing to bring things to me any time this would be more convenient.

We lived in a 1940s–era home on a street named Walkup. There were two steps, widely spaced, in a concrete walkway that led from the sidewalk to our front door. We had a mimosa tree in our front yard, one we had brought home from a visit to South Carolina, and it was in bloom the first time the general visited our house. He admired the delicate flowers and made a point of asking about the tree.

The house had an unfinished basement and I'd told General Koch early on that this was where I planned to set up my home workspace. The basement at one time had held a coal furnace—still common in Carbondale at the time—but we'd replaced this with a modern oil furnace and air conditioning system. The new installation used the original furnace pipes to circulate heat, the pipes culminating in flat vents flush with the oak floors.

The poorly lit basement turned out to be a dismal place to work. Our boys were small and shared a single room, which left us with a spare bedroom and I set up my workspace there. This is where I took the general on his first visit. He walked into the room with a load of reference material, looked around, and smiled broadly.

"This is nice," he said. "I've been worrying about you working in some dark, damp, dungeon–like space in the basement. But this is nice." The air conditioning system was throwing out a stream of cold air. "It's air conditioned," he said, as if surprised. "It's coming out right there," indicating a nearby floor vent. "Yes, this is nice."

As I already had learned, Oscar Koch had a cat-like curiosity about anything he hadn't investigated before. He asked how the floor vents worked with the two boys. Didn't they drop toys and things through them, down into the open pipes?

"All the time," I said.

"And you have to fish them out?"

"Yes, I do. And sometimes they're hard to reach."

He noticed my new Royal Ultronic typewriter, one of the first portable electrics, which I'd bought when I grew tired of going back to my office on campus at night to type finished draft copy. He wanted to know all about it. Did I like it? How, exactly did it work? Would I recommend it if he decided some day to get an electric typewriter?

I invited him to sit down and try it. He declined. My word was good enough for him, and anyway he was not likely to give up his trusty old manual machine.

And this was the way our session went, questions about the boys, how was Mary? Would it be easier for me if he carried stacks of files and other research materials to our house? Please let him know and he'd be glad to do it. And he was sincere, concerned about me, how he might make my task easier.

Such consideration was what I had come to expect from Oscar Koch. It simply was in his character to think of others first.

It would be an understatement to say I enjoyed these discussions with the general. The way he listened, his self-effacing humor, his openness, the authentic laugh he shared very easily—all these made the time I spent with him pleasant time. No matter our best intentions, there were instances when our conversations wandered far afield from the labors of research and writing and there always was something new to talk about. Often it was more bad news reflecting the times in which we lived.

The animosities between black and white Americans continued to worsen in 1967, and opposition to the war in Vietnam continued to grow. Dramatic happenings such as the inauguration of avowed segregationist Lester Maddox as governor of Georgia early in the year assured that strained race relations were not likely to improve any time soon. Tens of thousands of people were involved in race riots in San Francisco, New York, Minneapolis, Detroit, Newark, Washington, D.C., and other cities during the coming months. Rioting led to hundreds of injuries and a number of deaths.

The UPI wire service reported in October that Washington, D.C., had seen its biggest antiwar rally yet, with a massive but orderly crowd marching on the Pentagon shouting demands for America to get out of Vietnam. While police put the official estimate of the crowd at twenty-five thousand, the protest organizers claimed they had somewhere between 150,000 and 200,000 marchers.[2]

At the same time, 1967 saw important advances in the status of black citizens. The U.S. Supreme Court ruled that all state laws prohibiting interracial marriage were unconstitutional, and not long afterward Thurgood Marshall was confirmed as the Court's first African-American justice.

And old wounds from World War II still were open. SS Lieutenant General Wilhelm Harster was brought to trial. Head of the German security police during German occupation of the Netherlands, he was implicated in the murder of some one hundred thousand Jews. A Dutch court sentenced him to twelve years in prison, but he would be released after serving only three.³

We might take a few minutes to discuss these things when they happened, but we tried not to lose too much time from our work. Seeing our book through to completion still was our top priority. Our operative goal was "the sooner, the better," even if we didn't have an urgent deadline. But we soon would have.

Circumstances changed on a beautiful autumn afternoon I still remember vividly.

The day started well, except for the mimosa tree. It had been proliferate in flowering through a long hot summer but as I left the house I noticed that all its fragile blossoms finally had dropped off. I was to see General Koch in the late afternoon and, as always, looked forward to our session.

My workday on campus was routine. I caught up on a few small things that had been neglected for too long and took satisfaction in marking them off my "to do" list. At four o'clock I called Mary to remind her I'd be late and rushed to the parking lot and drove the few miles to the Koch home west of town.

As soon as I entered I sensed that something was wrong.

The general's face was unusually stern. We exchanged quick informal greetings and then he said abruptly, "I may as well get to the bad news first. I've been diagnosed with cancer."

I was stunned. I fumbled for a response. "I didn't know you'd been sick," I said weakly.

"Actually, I hadn't. I was standing with my arms folded

across my chest and felt a lump."

Perhaps it is this initial report that led me to believe that his problem was breast cancer, which may not be correct. Dr. Murray Zimmerman told me later the primary issue was gall bladder cancer, although it may have spread.

It was not easy for me to get accurate information. The general was quickly transported to Washington, D.C., for treatment and Nan held the old-fashioned view that to have cancer was, somehow, a personal disgrace and simply would not talk about it. Regardless of the specific details, though, General Koch was gravely ill.

The radio in my car almost always was set on a station that played rock music around the clock. When I started home that day, the Rolling Stones' new hit song, "Paint it, Black," was playing. The depressing lyrics added to my feelings of helplessness and despair.

I hurried home to tell Mary the bad news. She burst into tears at the word "cancer." I tried to comfort her—and myself, no doubt—by rationalizing that we didn't know how bad it was yet, that probably he could be cured and still enjoy a long life. In my heart I hoped this was true; in my mind, I had no such faith.

Over the next few months, there were relatively long periods during which the general and I had only sporadic contact. I used this time to read documents and make notations about things of interest, things that appeared to me to be important enough to be included in the book. I was not the one with the knowledge and background to make these decisions—although I already had learned a great deal—and hoped I would be able to discuss them with the general at some later point. Usually I was.

One bit that I stumbled onto made me particularly curious. The Seventh Army after-battle report on the Sicilian operation, in an accounting of G-2 Section activities, included a summary of the work of a unit listed as "Secret Intelligence (b) British." The entire entry was but a single sentence: "This unit, consisting of a party of five, also operated under higher headquarters and with experimental detachment G-3 (X-2), maintained close contact with the C.I.C. [Counter Intelligence Corps] and supplied valuable information."[4]

This was one I never got a chance to ask the general about. Perhaps that's good. It saved me from putting him on the spot. The five men in question were the British special liaison unit responsible for disseminating ULTRA intelligence to the Americans.

Although the history of the ULTRA project is well known today, at this time it still was secret. Its existence was first revealed in 1974 in the book, *The Ultra Secret*, by retired Royal Air Force Group Captain Frederick W. Winterbotham.[5] Captain Winterbotham praised Patton's effective use of ULTRA intelligence. I have little doubt that, had I asked about the British secret intelligence unit, the general would have told me just enough that I could understand in the broadest terms and know why its existence should not be revealed. If I didn't ask, he would not initiate a discussion of ULTRA.

Actually, he could have referred me to Ladislas Farago's *Patton: Ordeal and Triumph*, which was published a full decade before Winterbotham's book. While Farago doesn't use the ULTRA name, he refers to the British Signal Intelligence Service, which he says maintained "constant vigil" on the enemy's communications traffic and "translated" intercepted messages by breaking the Germans' "intricate codes and ciphers."

Farago's book cites the work of Major Melvin C. Helfers, the top ULTRA officer attached to Patton's Third Army, in the period before the Battle of the Bulge. He says, accurately, that Helfers had been monitoring the German radio traffic as usual and translating the intercepts as far as possible.[6]

As I would come to learn, there was a great deal more to this story. And Oscar Koch was highly instrumental in the successful use of ULTRA information in the Patton commands while being careful not to let it overshadow all the other sources. The lives of many American troops might have been saved had intelligence officers in high headquarters exercised the same care.

Chapter 10

PREPARATIONS FOR any large military operation take a staggering amount of effort. I had not fully grasped the complexity of General Koch's work until we got into the topic of planning for the July 1943 invasion of Sicily. Unlike the North African assault in which he had been Harmon's chief of staff, this time he was Patton's G–2 from the outset. He had a veritable treasure of original documents for me to dig into and he readily put everything at my disposal.

President Franklin D. Roosevelt and British Prime Minister Winston Churchill had met in January in the Casablanca suburb of Anfa and decided the Allies would invade Sicily as soon as operations in Tunisia were completed. Oscar Koch was sent back to the I Armored Corps in French Morocco, the earlier Patton command, to direct intelligence activities. Planning included reorganization of forces for what would become Operation Husky, the Sicilian invasion, scheduled to take place under "a favorable July moon." Once the invasion kicked off, Patton's force would be re–designated as the U.S. Seventh Army.

From an initial operating staff of three men—Koch, a

draftsman, and a stenographer—the G–2 Section within a few months would grow into one with hundreds of workers, including attached specialist teams. As we got deeper into the record, it became apparent that all of them were needed. And with the prodigious amount of work to be done, all of them surely were kept busy. Once again, Koch faced the perilous situation in which overlooking a single small, seemingly insignificant detail could lead to the deaths of American soldiers in combat.

"G–2 was the first section to be fully staffed," the general explained. "And it should be. Intelligence is the foundation for all the other planning." In this he echoed the declaration of Robert S. Allen that, "In planning, G–2 always had the first say."[1] Allen, of course, was speaking about the way things worked in Patton's Third Army while General Koch was stating a broad principle.

Early in the invasion planning process, Koch asked for photo interpretation officers and counter intelligence and prisoner of war interrogation personnel for assignment to the headquarters of lower units. Some two hundred interpreters were called for, based on the need for one in each battalion and more for regimental and division headquarters. This not only would speed up the questioning of captured enemy troops so that any actionable information uncovered could be put to use quickly, but it also would help in the winnowing out process. POWs with little or no information would not need to be sent on to higher headquarters for interrogation.

The general said a significant amount of the planning done for Sicily was based on lessons learned earlier. Experience had been an effective teacher, and Oscar Koch was a very good student.

"We'd learned in the Moroccan campaign that the G–2 Section needed a translation team," he recalled. "I considered this an essential addition to the intelligence staff. If we had information that needed to be translated, having to wait for translators could cost us time we couldn't afford." He requested and received an eight–man team of translators.

The first order of G–2 Section business was the preparation of a "Black Book" on Sicily. As General Koch explained it, the Black Book was a synopsis of material readily available in more detail elsewhere. "A ready reference, if you will," he said.

This compilation would include basic information such as a topographic study and comprehensive summaries of the island's fixed defenses, highways, railroads, airfields, port facilities, communications systems, beaches, and German and Italian forces. There also were special studies of local utilities, such as water and electricity, and other resources that might affect Patton's operations.

"And you had to compile this from scratch?" I asked, assuming I already knew the answer.

"Put it together, yes. But most of the material was already there somewhere. We just had to find it."

I protested that this sounded to me like a big job.

The general laughed. "This was the easy part," he said.

As we went further into his material, I could see that he was right. An intelligence officer, particularly a conscientious one like Koch, faces a truly daunting task.

One of Oscar Koch's first directives called for air photo reconnaissance, to be provided by a squadron based in North Africa. The intelligence staff would be particularly vigilant for any indication the enemy might be aware of the invasion plans. As an example, signs of railroad or bridge construction might indicate planned troop movement to reinforce the coastline defenses. Also of great value were the unique high oblique aerial photographs from flyovers at a mere two hundred feet altitude. These would be used in the construction of terrain relief models of the landing area beaches.[2]

Trained photo interpreters who analyzed the aerial photos were able to pick out the best landing sites. Information was so detailed that even the type of sand was known. Beach exits, underwater barbed-wire barriers, and defense installations were visible, providing details that were crucial to the invasion planning and no doubt saved lives.

"This was a big step up from the Morocco landings," the general said. "For that one, we had to use old picture postcards to find the best landing site near Safi." He chuckled, and added, "I know. It sounds like an amateur operation."

I said, "You're not serious."

"I am serious. The Navy had systematically collected all the beach and waterfront pictures it could get its hands on and had them cataloged and filed in Washington. We found postcards that showed us what we needed to know. Never underestimate a good intelligence team's resourcefulness."

"Especially an Oscar Koch intelligence team. You knew what you needed and found a way to get it."

"Well, I wasn't G–2 for that operation."

"You were chief of staff. You called the shots."

He nodded, as if affirming that I was beginning to catch on. And I was.

Aerial photo coverage continued to be of immeasurable

value throughout the Sicilian campaign. In the Seventh Army after-battle report, Koch doubled down on the need for extensive air reconnaissance: "Some logical positions of defense (based on terrain studies) were flown twice daily—for comparison. This procedure must always be emphasized as there is a tendency to consider such action uneconomical, but the balance of cost versus saving of American soldiers' lives gives the answer."[3] As always, his standard measure of success was the number of lives saved. And, as always, Oscar Koch was adamant that no price could be put on this.

Among other sources, Koch turned to the fledgling Office of Strategic Services—the OSS, forerunner to the Central Intelligence Agency—for help. The OSS had busily recruited Italian-Americans specifically for behind-the-lines operations in Sicily and, later, on the Italian boot. Requests had to go through higher headquarters, as the OSS at this time took directions only from the top level of Allied command.[4]

The general had great respect for OSS operatives. He called on the agency only for specific information difficult to obtain through the more conventional channels.

"They rarely if ever failed to get what we asked for," he said. "We never knew, and never wanted to know, how they got it. Sometimes they parachuted agents in from bombers, going out through the bomb-bay doors with the bombs. 'Stout fellows,' as the British would say."

Later in the war, each army had a special OSS intelligence detachment. Koch and the Third Army detachment had an excellent relationship. The Third Army OSS detachment completed more than one hundred missions behind the German lines and provided invaluable intelligence information.[5]

As planning for the Sicily landings intensified, Koch was directed to submit a request for the personnel he needed in the G–2 Section. He asked for enough officers to bring the staff total from the existing 85 to 178. Included in the directive was permission to request individuals from other units by name, which was a generous allowance he took full advantage of.[6] Such authority from higher headquarters clearly demonstrated the critical nature of his role as intelligence chief as well as the trust placed in his judgment.

On a purely functional basis, little things meant a lot. Koch had developed a system of clear acetate overlays that would allow commanders to superimpose sheets displaying elements such as maps, terrain features, and the disposition of enemy troops, one on top of the other. The essential transparent sheets always seemed to be in short supply.

"We seldom got as many as we ordered, and everyone on the G–2 staff was encouraged to find creative ways to find more," he explained. "Sometimes we even used discarded x–ray film from our hospitals." Then, with a guilty smile, "If extra sheets happened to show up, nobody asked where they came from."

Robert S. Allen recalled that at one point Koch went so far as to impose a one–dollar fine on any member of the G–2 Section who posted one of the sheets with a thumbtack and left a hole that damaged the acetate.[7] No matter the system was quite simple, the transparencies were much too valuable to be handled carelessly.

As an example of the detail necessary for a comprehensive invasion plan, although it did not involve intelligence, the Personnel Section had to estimate the number of medals that might be needed and order half this number in advance. The initial order requested, among others, eight Medals of Honor and fifteen thousand Purple Hearts.

Meanwhile, the Medical Section requested and was granted the G–2 Section's studies of the island to help it prepare for special medical problems that might develop during the operation. It made lists of hospitals, as well as schools and hotels that might be used as emergency hospitals, and diseases peculiar to the island.[8]

Because the invasion would be a joint American–British operation, Koch issued a directive that intelligence staffs at lower levels be trained to understand British intelligence and tactical practices, particularly with reference to abbreviations, nomenclature, traditional signs and symbols, and the standard British grid system used for mapping. He said there was no room for miscommunication among friends, "and the Brits always had their own way of doing things."

Along with everything else, invasion planners had to prepare for the use of new equipment.

"We were going to get a lot of Dukws [amphibious trucks] and we didn't have very many soldiers who knew how to drive one," the general said. Potential Dukw drivers and maintenance men were sent to a special engineer brigade for training.

In after–battle reports, the Dukws received rave reviews. The awkward–looking vehicles played an essential role in later landings, as well. The journalist Ernie Pyle writes that Dukws were a crucial element in keeping supply lines open. He says that after the invasion of the Italian mainland at Anzio, "Day and night, a thin, black line of the little boats moved constantly back and forth between shore and ships at anchor a mile or two out. They reminded me of ants at work." He says there were "hundreds" of the amphibious trucks in use at the site, each with only one crew member, the driver.[9]

"You know what a Dukw is, of course," General Koch said. It sounded like a question.

I told him that not only did I know what a Dukw was, but that I actually had a Dukw story of my own. He wanted to hear it. I always got the impression he welcomed interruptions for personal conversation not related to our work, even though we both were eager to move the project forward as fast as we reasonably could. The problem was, one aside typically led to another. If we weren't careful, such talk could eat up an afternoon.

"It really isn't important," I assured him.

"Doesn't matter. I'd like to hear it."

I told him about a Fort Jackson open house at which one of the featured attractions was Dukw rides on a small lake in the central area of the post. At their mother's request, I took two young boys to the event. Their names were Jamie and John and they lived next door to Mary Corley, by this time my fiancée.

The boys and I spent a long afternoon watching paratroop landings and looking over tanks and artillery pieces and other heavy weapons of war that rarely were seen at Fort Jackson. It was getting late. I was ready to take them home, but they insisted on a Dukw ride first. They said I'd promised, and maybe I had.

We got to the lake and there was a long line of people waiting for rides.

"But you waited," the general said, laughing.

I told him yes, we waited ... and waited. I was afraid the boys' mother would be getting worried. And I knew Mary would. And I was right. By the time I got the boys home I was in trouble. Jamie and John came through like troopers. It was their fault, they proclaimed loudly. I was great for letting them wait. They got to ride on a Dukw and it was even more fun than they expected and they knew their mother wouldn't have wanted them to miss a chance like that! And neither would Mary, right?

Thanks to them I was off the hook. I wanted to applaud their performance.

"Heroes are like intelligence officers," General Koch said. "They're made, not born."

"And sometimes they get lucky," I said. I hoped, later, that he didn't take my comment to mean he had been lucky. But the general didn't play those kinds of games, and I need not have worried. He probably would have been the first to admit that yes, sometimes he had been lucky. I would have argued that he made his own luck.

As I knew very well, the philosophy he'd just stated, albeit facetiously, was a driving force in his determination that intelligence specialists needed to be trained before they hit the ground in combat. The front lines were a good place for trained intelligence officers to learn more through experience, but on–the–job training was a luxury best avoided.

In Oscar Koch's own case, though, much of what he knew about combat intelligence in fact had been learned through personal experience. Planning for Sicily may have added an even thicker layer of new knowledge than he wanted. Higher headquarters revised the original invasion plan, and much of Koch's work had to be revised as well.

And there also was a critical new factor: as combat in North Africa wound down, German forces there were likely to be relocated to Sicily. Early accounting of German troops on the island could be thrown into a cocked hat.

The output of the G–2 Section during the planning stage of the Sicilian operation was immense. Koch's first G–2 Estimate of the Enemy Situation was issued on April 20, almost three months before the prospective invasion date. It estimated a total Italian garrison of 153,000 troops on the island, seventy–five thousand of these assigned to coastal defense units, and anticipated that some twenty thousand German troops would be evacuated from Tunisia to Sicily by the time the Allied troops landed.

Tight security was essential throughout the planning process, and nowhere more so than in the G–2 Section. Koch was keenly aware of the damage leaked information could do.

As a backstop, just in case an intelligence report fell into the wrong hands, he always gave the enemy situation not only as it applied to Sicily, but also to southern Italy, Sardinia, and Corsica. This paralleled an extensive counter–intelligence plan that had Patton visit Corsica and other potential invasion sites in an effort to mislead enemy intelligence should anyone be watching.

Estimate No. 2, issued two weeks later, upped the number of troops on Sicily to 165,000. Koch now predicted the Allies would face six Italian divisions, two German divisions, and 84 coastal defense battalions. Details of enemy unit identification and location were given, along with estimates of their capabilities. The "favored" capability was that the enemy would defend at the water's edge and counterattack to the east from a general line from Licata to Ravanusa.

Koch believed the seaport and four airfields in the Gela area where the invasion would be launched were of great strategic importance to the enemy and could only be defended adequately at the coastline. He said the counterattack from the northwest was logical because of the enemy's heavy concentration of mobile combat units in the area around Caltanisseta. "The terrain lends itself to counterattack from the northwest and the rivers would not form a serious obstacle at this time of year because they are reported dry," his report advised.[10]

A new estimate issued on June 9 identified Italian military units on the island that, if at full strength, would total 208,500 troops. Further, it stated that a "considerable potential of reserve manpower" was available. The report said invasion troops might face six or seven Italian divisions, two German divisions, and the equivalent of six coastal divisions. It estimated the enemy would have 682 "serviceable" German and Italian aircraft available to attack invading forces.[11] And Koch still expected the enemy to defend at the water's edge.

The general said a number of high–ranking visitors from the States did not agree with this latter conclusion, which I found surprising. The group included representatives of General George C. Marshall, the Army chief of staff back in Washington. When they visited Seventh Army headquarters, Koch gave them a detailed briefing complete with maps and charts for graphic support—the kind of briefing at which he was greatly skilled.

He closed with a discussion of the enemy capabilities and ended with a prediction the enemy forces would defend at the water's edge, consistent with the position he'd held from the early days of invasion planning. Then he learned very quickly that his view was not universally accepted. General Albert Wedemeyer, who was present as one of General Marshall's personal representatives, stayed after the others had left the G–2 Section briefing room.

General Koch was quite specific in explaining what happened next, and I wondered how it might have felt to be told that everyone else thought you were wrong.

"General Wedemeyer told me I was the only one who thought the landings would meet major defense at the water's edge," Oscar Koch reported. "He said, 'The War Department doesn't agree, AFHQ [Allied Force Headquarters] doesn't agree, other headquarters don't agree, and I don't agree.' He went on to say that all these folks believed the enemy would defend inland against the Americans and counterattack against the British to the east."

"Why?" I asked, making no effort to hide my sarcasm. "Did he think they knew something you didn't?"

"His explanation was that the Italians' and Germans' main goal would be to protect the Catania Plain with its airfields and hold onto the road to Messina."

"Did that make sense?"

"It made some sense. And it actually would be easier on us if they did. But I had full confidence in our intelligence and stuck to my guns. Wedemeyer left the room insisting I was wrong. He was pleasant about it, though, and I took it in good spirit."

Another story for the book. This is the way we told it in *G–2: Intelligence for Patton*:

> Early on the morning of July 11 [the day after the Allied landings], a major counterattack was launched against the Gela beachhead.

> Twenty German Mark IV tanks approached to within 2,000 yards of Gela before being stopped by artillery. At about 4:30 P.M. they renewed the engagement in another vicious counterattack. This time American forces had to call upon all means at hand—including naval gunfire, then a novelty against ground troops—to repel the advancing enemy force. But the enemy attack again fell short....
>
> The enemy had tried unsuccessfully to defend at the water's edge as we had predicted. General Wedemeyer graciously stopped by the advance command post at his first brief respite. He visited only momentarily, just long enough to tell me, "You were right. The others were wrong."[12]

As military invasions go, Seventh Army's landings on Sicily would have to be classified as successful. American troops took the enemy by surprise and suffered a minimum of casualties. The enemy's coastal divisions were trained to defend the coastline, to prevent landings if they could or at least to delay the invaders until mobile units inland could arrive in sufficient force to launch a counterattack. But no significant mobile forces had come to their aid. The Americans soon had a solid footing on the island from which to push northward to the ultimate major objectives of Palermo and Messina. Oscar Koch's intelligence work had been superb.

Two days after the landing, Patton's headquarters was moved inland to a school building in Gela. By the following day, it became clear the enemy—his desperate but limited counterattacks proving costly failures—was beginning to withdraw. But Sicily was yet to be taken.

Chapter 11

ONE OF THE things General Koch admired and respected most about Patton was the way he backed his staff, commonly known in military circles as command support. He said Patton had faith in his staff and let staff members do their jobs without interference. Needless to say, this was contingent on the staff officer being competent and doing his job well. This never was an issue in Oscar Koch's case, certainly, and Patton was never sparing in his praise for Koch's work.

There is an oft–cited Patton quotation that labeled Oscar Koch "the best damned intelligence officer in any United States Army Command."[1]

Patton believed strongly in the value of loyalty from the top down. He claimed that it is "this loyalty from the top to the bottom which binds juniors to their seniors with the strength of steel." One of his stated principles of leadership was to "always give credit where it is due. If you win, give the credit. If you lose, take the blame."[2]

General Koch said his old commander "claimed to be more interested in a loyal staff than a brilliant one." He said Patton never needed to worry about his staff being loyal, and his staff was grateful that with him it was a two-way street. Visitors were amazed by the faith the commander showed in his staff and junior officers. "Sometimes he acted like a proud parent when we made reports and other commanders were present. And he always thanked us after we'd finished."

G–2 was a critical staff position and Patton kept Oscar Koch close at hand. No big decisions would be made without consulting his intelligence chief.

Given all this, General Koch wanted to begin our book with an example that illustrated definitively not only the way Patton used intelligence but also the extent to which he trusted his intelligence staff. We used an instance from the U.S. Seventh Army's sweep across Sicily, six days after the Allied troops hit the beaches.

Patton asked him a simple, direct question: "If I attack Agrigento, will I bring on a major engagement?"

Koch's answer was equally direct: "No, Sir."

Patton directed Colonel Halley G. Maddox, the Seventh Army operations officer (G–3) to order the attack.

The general would write later, in defining command support, "In Patton's commands, intelligence was always viewed as big business and treated accordingly. Although working, by necessity, in the shadows, it always had its place in the sun. It was never viewed as subordinate to any other staff activity. G–2 was never the forgotten man."[3]

I would have preferred to use that "big I" personal pronoun instead of "G–2" in this last sentence. But General Koch resisted my efforts. His modesty was always a factor, but I also think that through his years of service with Patton he virtually had come to identify himself as "G–2." In the military tradition in which Koch was well steeped, it was common for other Patton staff members to refer to him this way, as in "Let's check it with G–2."

Like me, Oscar Koch's cousin, Dr. Murray Zimmerman, was bothered that Koch didn't give himself more credit. "I'm sorry that Oscar characteristically said 'G–2 did this' or 'intelligence did this,'" he said. "I wanted him to say 'I fought like hell, and I was the only one to say this. It was not intelligence or G–2, it was me!'"[4]

Shortly after the fighting in Europe ended, Dr. Zimmerman, a Medical Corps captain, witnessed the congeniality common in Patton's Third Army headquarters. Like most visitors, he was taken by surprise. In describing his experience, he coincidentally substantiated the prominence typically accorded G–2.

Captain Zimmerman recently had been transferred to a Third Army tank battalion. He had been reluctant to call Koch before, during combat, but decided it was time to get in touch.

As he recounted the story, "I got through to headquarters and he said, 'Give me your coordinates on a 1:500,000 [scale map] and I'll have a plane up there for you in a little over an hour.' I went to my colonel and told him I was leaving for the weekend, and when he said 'No way!' I explained that my cousin, the G–2, was sending a plane up to pick me up and would probably be a little cross if I didn't come back on it. He smiled benignly and said, 'Have a good time!'"

Dr. Zimmerman recalled a marvelous dinner with Patton and the Third Army staff.

"Patton was very laudatory about Oscar. I remember well that Patton told me Oscar was the only one on his staff who knew that Regensberg and Ratisbon were the same thing." (Regensberg, a German city situated on the Rhine

River, was the scene of one of Napoleon's victories. Known by the French as Ratisbon, it was made famous by the Robert Browning poem, "An Incident of the French Camp.")

Patton also told Captain Zimmerman about something that took place during Third Army's drive on Germany, when he and Koch and General Hobart Gay, Third Army chief of staff, along with a driver, had gone forward to a point near the front lines one day to "see what was happening on the ground." They were standing in an open place on a knoll overlooking a narrow river and caught a glimpse of a German soldier on the other side just as he disappeared into the trees.

After a few minutes of observation, Koch checked his watch and said, "It'll take two minutes for him to get back to his headquarters, another two minutes for them to call regimental headquarters, and then two more minutes for the regimental artillery to focus in on this grid square. I think we should get the hell out of here." Right after they left, the place where they'd just been standing came under heavy shelling.

"Oscar probably wouldn't have said 'the hell,'" Dr. Zimmerman added, "but that's the way Patton told it."[5]

Although General Koch gave Patton much of the credit for the command support his staff enjoyed, it's clear that Patton was not a commander who showed great tolerance if a staff officer made an error. As Koch remembered him, Patton never wanted to be burdened with needless detail. That was staff work. He expected his operational staff to give him information he could depend on. This was true in spades when it came to intelligence.

When the decision concerned launching a major attack, as was the case when he asked about attacking Agrigento,

Patton knew errors would be measured in terms of lives lost. He also had full confidence in his intelligence chief; Oscar Koch would give him the right answer.

General Koch pointed out that, as with most intelligence prognoses, his quick, direct response to Patton climaxed months of cumulative effort. All the available sources of information had been used, bits and pieces had been fitted together like pieces of a gigantic jigsaw puzzle until a clear picture emerged. Given all that was known, the Germans and Italians clearly were not capable of putting up a strong defense or launching a counterattack if Patton struck their forces at Agrigento.

"You knew all about them," I said.

"That's what I was there for."

He made it sound simple.

As we talked about Sicily one day, I asked the general about the infamous slapping incident, in which Patton struck a soldier he felt was malingering. He couldn't—or wouldn't—tell me much beyond what already was well known. He didn't witness it and, not surprisingly, it was not a subject of conversation around Seventh Army headquarters.

He believed that Patton had been driving himself too hard at the time and was extremely tired. Although Patton apparently refused to acknowledge the existence of combat fatigue under any label, it always was expected that a number of troops in an invasion force would experience serious emotional issues. In the case of Sicily specifically, U.S. Seventh Army planners estimated that neuropsychiatric cases would account for 15 to twenty percent of all non–fatal battle casualties during the first few days of combat. The standard treatment prescribed was heavy sedation and evacuation as an "ordinary casualty."

Medical personnel aboard ship were encouraged to administer phenobarbital to the troops two to five hours before the landing. Not only would this "materially decrease seasickness," but it also would "decrease natural but excessive nerve tension."[6]

Seventh Army after–battle reports showed that 702 hospital admissions during the Sicilian campaign were because of neuropsychiatric problems. Of these, 386 soldiers were too serious to return to duty and were evacuated from the combat zone. In the El Guettar battle in Tunisia, where relatively green American troops were involved in heavy fighting, twenty percent of the total combat casualties were classified as neuropsychiatric.

In the Lessons Learned section of the Seventh Army after–battle report, a sub–section on mental attitude stressed that "enthusiastic elan and self–confidence" on the part of troops making an initial assault was of greatest importance. It went on to warn that, "Battle is a violent and elemental occupation. Men to conquer in battle, particularly in a night landing, must be imbued with this elemental viciousness." Fierce warriors were in demand.[7]

Seventh Army Headquarters was aboard the transport ship *U.S.S. Monrovia*, which later would be transformed into a hospital ship. As soon as the invasion fleet set sail on the Mediterranean, a letter from Patton was released to the troops. It concluded:

> Above all else remember that we as the attackers have the initiative. We know exactly what we are going to do, while the enemy is ignorant of our intentions and can only parry

our blows. We must retain this tremendous advantage by always attacking; rapidly, ruthlessly, viciously, and without rest. However tired and hungry you may be, the enemy will be more tired and more hungry—Keep punching! No man is beaten until he thinks he is. Our enemy knows that his cause is hopeless.

The fact that we are operating in enemy territory does not permit us to forget our American tradition of respect for private property, non-combatants, and women....

The glory of American arms, the honor of our country, the future of the whole world rests in your individual hands. See to it that you are worthy of this great trust. God is with us. We shall win.[8]

As ships loaded with the men who would go in over the beaches neared the Sicilian coast, C–47 Dakota transport planes took off from Kairouan, Tunisia, carrying elements of the 82nd Airborne Division that were to lead the attack. Meanwhile, the Navy had opened up heavy shelling on enemy installations near the beaches to soften the German and Italian defenses.

But not everything went as planned.

"Patton always said anything made by man could be destroyed by man," General Koch said. "And I suppose you might say the other side of that coin was the element man could not control—nature. You always try to figure the weather into intelligence calculations, and the Sicily landings offer a good example why."

I asked him to explain.

First of all, the general said, as the invasion fleet crossed the Mediterranean it ran into a violent storm. The landing almost was called off, to be delayed for at least a day. This proved not to be necessary in the end, but the rough seas led to a large number of the American troops being stricken by debilitating seasickness. Nonetheless, they would be hitting the beach near Gela in the black of night.

The paratroops' target drop zone was an area of high ground northeast of the Ponte Olivo airfield, some six miles inland. High winds had forced the C–47s off course, though, and the airborne troops were scattered over a large area of southern Sicily. One battalion landed completely out of the American zone of attack but regrouped quickly and fought alongside British units for the next couple of days.

I waited until the general had finished his detailed account. There was something I'd always wanted to tell him and this seemed to be a good time.

"I wanted to be a paratrooper," I said. "I actually applied for airborne."

"Did you really?"

"Yes, I did. Just because I wanted to ride in airplanes."

"I take it you didn't make it," the general said. "What went wrong?"

I told him I couldn't pass the demanding physical examination necessary for airborne training. This required a "picket fence" physical profile (all ones), and I had a couple of twos. I had slightly flat feet and I already had some disk problems in my spine. "So the Army decided I was better suited for administrative school," I added.

"I'd count my blessings," he said. "Jumping out of a C–

47 and coming down in enemy territory in the dark never seemed to me like the best way to go into combat."

I agreed, although the scenario he described actually didn't sound much worse to me than jumping into the ocean with a heavy pack and trying to make it to shore before drowning or getting shot. But this sentiment was one I supposed didn't need to be expressed.

Anyway, we needed to turn the discussion back to the 82nd Airborne Division and the invasion of Sicily. And I didn't particularly care to spend more time talking about my very undramatic military service with a man who looked more heroic to me every day.

"Those paratroopers were tough," I ventured.

"The men of the 82nd were as good as any we had. It wasn't their fault they were lost before they ever jumped. But having that first wave of troops miss their targets was a discouraging way to begin an operation. Patton had written the soldiers that God was with us, but I'd guess that a lot of the 82nd boys coming down in the dark with no idea where they'd be landing might not have felt too sure about that."

He laughed at his own commentary, then sobered. "It's easy to laugh now, but nobody was laughing that night. I've always thought a seaborne invasion is one of the most dangerous things an army could ever experience."

"How does it feel back on board the ship?"

"It's painful. You want to be out there in the water with them. You're scared for every single man. But pretty soon you're too busy to think about it. Things begin happening pretty fast."

The Seventh Army assault troops were in landing craft moving toward the beaches shortly after midnight. The

Italian defenders clearly were taken by surprise. There was no large-scale resistance to the first wave of troops and all the landings had been accomplished by 6 a.m.

Enemy reaction to the attack was not slow, however. Landing craft unloading reinforcements and supplies on the beaches soon were under attack from the air, being bombed and strafed all along the Seventh Army zone of attack. By 8:30 a.m., Koch's G-2 Section had reports of enemy tanks moving toward American forces. Allied aircraft were called in for quick strikes. By early afternoon, large numbers of Italian troops, including armored units, were known to be on their way, presaging the heavy fighting to come.

By the end of the day, Patton's troops had secured all of the invasion beachheads and had advanced inland as far as four miles in some areas. The scattered 82nd Airborne units had begun to make contact and reform, at least two battalions engaged in heavy combat by mid-day. More than four thousand prisoners of war had been captured.

Success had not come without a price. American casualties, from both enemy fire and land mines along the beaches, amounted to 58 killed, seven hundred missing in action, and almost two hundred sick and wounded. Half of the latter were taken aboard hospital ships and a civilian hospital in Gela was requisitioned to take care of the rest.[9]

"As it turned out," General Koch said, "the first day of the invasion was just the calm before the storm. The Germans and Italians took a day to regroup and early the next morning they hit us with everything they had. But nobody had ever expected the invasion of Sicily to be easy. We expected a fight and we got one."

Given the successful results of dropping propaganda

leaflets on enemy troops in Tunisia, Koch expanded this activity in Sicily. Fifteen million leaflets were dropped during the first week of the campaign. A psychological warfare detachment in the G–2 Section targeted propaganda messages based on information gathered from POW interrogations and captured documents and mail. While the large drops continued to originate in North Africa, some half-million localized leaflets were printed in Sicily and dropped from Piper Cub planes or shot into enemy lines by artillery shells. In several instances these contained enemy commanders' names gained from prisoner interrogation.

The leaflets proved highly successful in leading enemy troops, especially the Italians, to surrender. Among Italian POWs questioned, some eighty percent either carried one of the leaflets or said they were familiar with their contents.

Not all the propaganda was aimed at enemy troops. The G–2 Section estimated that in a number of cases the surrender of Sicilian towns was hastened by the scattering of leaflets aimed at civilians. The residents were forewarned that their towns were immediate military objectives of the Americans and urged to fly white flags of surrender to avoid further damage to their homes and businesses. The general said it was hoped this effort would save both lives and unnecessary destruction of property and help create goodwill among the Sicilians.

Propaganda programs also were broadcast by radio, first from North Africa and later from captured Sicilian stations. Movie films were seized and censored, and replaced in some cases by movies with American and British propaganda value. The G–2 psychological warfare detachment also started a daily newspaper, *Sicilia Liberata*. It began publication in early August, using the staffs and production facilities of two Sicilian newspapers. Detachment photographers took pictures of American troops being welcomed as liberators by local citizens.

"Were they responsible for any kind of subversive action?" I asked.

"As far as the intelligence section was concerned," the general said, "the only thing was rumors spread by the psychological warfare branch."

"Rumors?"

"Yes. Things like the reports of fighting between German and Italian soldiers. We knew all along that the tie between the Italians and the Germans was pretty flimsy. Anything we could do to weaken it further seemed like a chance worth taking."

He said the G–2 Section psychological warfare detachment also made contacts with anti–fascist organizations throughout the island. These, in turn, helped identify possible friendly civilians and led to valuable intelligence information.

I no longer was surprised when new sources of information tapped by Koch's G–2 Section were mentioned. I wasn't familiar then with the definitive label "all–source intelligence," but when I heard or read it the first time it struck me as the perfect tag for Oscar Koch's method. No potential source of information that might fit somewhere in the intelligence mix was overlooked. By the numbers, the principles of intelligence operation began with "collection."

Chapter 12

SOMETIMES, IN COMBAT, critical information comes from the most unexpected sources. On the third day of the Sicilian campaign, an American officer who had been captured by the Germans escaped and brought confirmation of what Koch already suspected based on other intelligence at hand. There were two German divisions on the island, the Hermann Göering Division and the 15th Panzer Division. In addition, there were four companies of German Mark VI tanks.

"We expected the Germans to put up more of a fight than the Italians," the general said. "Many of the Italian prisoners hated the Germans, and said they'd been deliberately sacrificed to cover the Germans' withdrawal."

In at least one instance, their complaint was borne out with the discovery that retreating German units had planted mines in the road as they moved, ahead of Italian units coming behind. This left the Italians trapped between advancing American forces to the front and mine fields to the rear.

Meanwhile, the Americans had gained a new ally of their own. The 4th Tabor of Goums from the French North

African Army landed on Sicily and joined the Seventh Army force. The French unit was attached to the 3rd Division initially and later the II Corps. It fought with the Americans through some of the most rugged terrain of northern Sicily. General Koch lauded the French soldiers, attesting to the report they "gave a good account of themselves through the whole Sicilian campaign."

As Koch's pre–invasion intelligence estimates had forecast, the Seventh Army troops had to fight from city to city and capture each one as they advanced. The paucity of good roads made it impossible for the cities to be bypassed. Gela, Licata, Comiso, and Ragusa soon were in American hands. Another two thousand prisoners were taken on the third day of battle, and accommodating the growing number of captured and deserting enemy troops began to be a serious problem.

Medical care for civilians in the region also posed a substantial challenge. The Army, given its limited medical supplies, could only render first aid to those in danger of dying without it.

In the first six days of fighting, Koch's intelligence reports estimated, 15,500 prisoners of war had been captured. The enemy had lost nineteen hundred dead and wounded, and hundreds of aircraft, tanks, and heavy artillery pieces had been destroyed. Total strength of the Seventh Army, with much lighter losses, still stood at more than two hundred thousand.[1]

An OSS detachment, which was made up of three officers and nine enlisted men, joined the G–2 Section on July 15. Koch knew exactly how he wanted to use them.

"We were looking ahead," the general explained. "We planned to have them infiltrate enemy lines in the Palermo area, but we moved so fast they never had a chance to pull this off."

I said, "With Patton, you never slowed down, right?"

"Not unless we were forced to, like Third Army was later when General Eisenhower shut off our fuel supply. But we'll get to that. We're not out of Sicily yet."

I was grateful to him for keeping us on track. With his firm hand on the wheel, I wasn't able to take off on wild tangents every time he mentioned something I was eager to know more about. We had agreed on a chronological scheme for the book, which I had favored, and thanks to the general we'd stuck to it. At this point there still was much to cover on Patton's Seventh Army operations in Sicily.

Agrigento fell to troops of the 3rd Division on July 17, just a week after the Allies came ashore. Koch's intelligence proved correct. Agrigento hardly slowed down the American forces as the Seventh Army fought its way across western Sicily. From Koch's contemporary reports, it is clear the Italian forces were in a state of great disarray. His G–2 Periodic Report on the 17th included information from a prisoner of war, a brigadier general who was commander of the Italian 207th Coastal Division:

> This General stated he had been charged with planning and execution of defenses of CASTELVETRANO airfield, which, due to our bombing, is now unserviceable. He was placed in command of the 26th Assieta Division 10 July before assuming command of the coastal division. He stated that great confusion exists in

the High Command with changes of commanders, and, when captured, he had on his person complete Battle Order of 26th Division, which he had prepared for his own information.[2]

With the German and Italian troops on the run, a new directive from higher headquarters assigned Patton a primary objective of Palermo. Meanwhile, the British Eighth Army under Allied Forces Commander General Sir Bernard L. Montgomery was ordered to drive the enemy on the other side of the island north and east into the Messina Peninsula.

Over the next ten days, the total number of POWs taken by Seventh Army rose to an estimated 52 thousand, with another seven thousand dead or wounded. The enemy's weapons of war were falling rapidly into American hands, also.

Captured aircraft and heavy artillery pieces totaled more than 220 each, along with more than 150 tanks.[3]

As the 3rd Division and the 2nd Armored Division moved close to Palermo, a coordinated attack was planned but no resistance materialized. An American commander received the formal surrender of the city on the evening of July 22.

"Up to this point, it had been almost too easy," General Koch observed. "We knew this was too good to last. The Germans weren't going to be pushed off the island without putting up some stiff opposition. After Palermo, they stopped running and started digging in."

Koch's G–2 Periodic Report No. 19, issued just after the fall of Palermo, included a succinct assessment of what

Seventh Army troops might face in the days to come. The enemy was in a position to stage a final defense along a line extending through Troina, the report noted, and while the Americans' surprise attack and swift advance had left the enemy little time to organize strong defensive positions in the beginning, conditions had changed. When this line was reached Patton's Seventh Army forces could expect fierce resistance.[4]

His projections proved to be right on target. Not only did the German and Italian forces already on Sicily begin an all-out defensive effort to slow the Allied advance, but air reconnaissance showed reinforcements crossing the Strait of Messina from the Italian mainland. The enemy was forming a solid line of defense between the Seventh Army and Messina, the ultimate objective of the American and British armies.

There also was some good news. Word of the fall of Benito Mussolini on July 25 quickly spread among the American troops. Now they had the satisfaction of knowing their assault on Sicily had helped to bring the first wide crack in the Axis political alliance.

"Of course we already knew that Mussolini and the Fascists had little popular support on the island," the general told me. "We'd captured five Italian generals and one admiral, and every one expressed disgust for Mussolini and his government."

He said the civilian populations of the captured towns, without exception, had greeted the American soldiers as liberators. Koch's G–2 Section had taken full advantage of this for its propaganda value. The Italian soldiers in the field could become discouraged quickly if it was apparent the people back home were not behind them.

Results were seen promptly. A POW captured July 27 claimed that twenty Italian soldiers had been shot for refusing to fight. A few days later, another prisoner said that fully half of the soldiers who made up the Sicilian Aosta Division had deserted and gone home.[5]

Meanwhile, Oscar Koch did a remarkable job of keeping ahead of the fast-moving advance by Patton's U.S. Seventh Army forces. At his request, the OSS detachment, working in three-man teams, penetrated enemy lines and provided much useful intelligence information. Even a mission that fell on hard times was productive in the end.

A team made up of the detachment commander, four enlisted men, and two civilians slipping into enemy-held territory in the direction of Messina was ambushed and captured by Germans troops. Some were wounded.

"These guys were hard to hold, I guess," General Koch said. "All of them except the detachment commander eventually escaped and made it back through the enemy lines with information that was highly valuable to our troops."

Had I been telling this to the general, his immediate response likely would have been, "What happened to the detachment commander?" With the situation reversed, though, I was too slow to think of that important question before moving on to other things.

In subsequent intelligence reports, Koch continued to emphasize the buildup of German forces along the defensive line noted. Coming events would prove the accuracy of his calculations. Seventh Army troops approaching Troina on the first day of August encountered the strong resistance he had been warning of. In a five-day battle, the Germans launched no fewer than 24 unsuccessful counterattacks. San Fratello was contested almost as bitterly, and with its fall on August 8 the final German retreat and withdrawal to the Italian mainland began in earnest.

"Troina blocked us from the limited road network they were using to withdraw to Messina," the general said, pointing out the town's strategic location on a map. "We could see that the Italians were being used to defend against the German escape to Messina."

Even as the hard fighting ended, the Americans faced slow going. The Germans blew bridges and laced the area with minefields as they retreated.

Nighttime evacuation of German troops continued. By mid–August, Koch's reports told the end of the story for the Sicilian campaign. No further defense of the island was indicated, as the Germans were frantically moving their troops across the Strait of Messina to the Italian mainland during the hours of darkness.

The conquest of Sicily had come at a cost to the Seventh Army of 1,233 killed, 4,695 wounded, and 968 missing. For the enemy, the cost was much higher: 100 thousand prisoners of war and an estimated twelve thousand killed and wounded. Among those taken prisoner were 33,700 Italian enlisted troops whose homes were in Sicily. They all were paroled almost immediately, along with 308 non–combatant chaplains and medical doctors.[6]

The Germans came out better than they should have, however, successfully evacuating a large share of their forces to the Italian mainland. The Seventh Army after–battle report downplayed this as "a small measure of success" for the enemy.

Italy, it predicted, would not be a safe refuge. Regardless, the fall of Sicily "ended forever the enemy's dreams of domination of the Mediterranean" and "the 'handwriting on the wall' had been seen by the Italian people, as it would soon be seen by other nations of occupied Europe."[7]

At the close of the successful campaign for Sicily, Patton issued U.S. Seventh Army General Order 18. It was, in typical Patton fashion, a glowing letter of acknowledgment to his troops. The commanding general's order was addressed to the "Soldiers of the Seventh Army":

> Born at Sea, baptized by blood, and crowned with victory in the course of 38 days of incessant battle and unceasing labor, you have added a glorious chapter to the history of war. Pitted against the best the Germans and Italians could offer, you have been unfailingly successful. The rapidity of your dash, which culminated in the capture of Palermo, was equalled by the dogged tenacity with which you stormed Troina and captured Messina.[8]

The commander closed his letter with a typically Pattonesque refrain, signifying his usual pride in the performance of his troops and his gratitude for their accomplishments in the conquest of Sicily: "Your fame will never die."

Chapter 13

OSCAR KOCH'S cancer diagnosis came as a frightful shock. It was painful to see a man with his vitality, suddenly and with no previous signs of illness, find himself in a life-threatening situation. His doctors sent him to the Army's Walter Reed Hospital in Washington, D.C., where he underwent surgery and cobalt treatment. He wrote me a brave little note saying his medical team had told him his treatment initially would aggravate his condition rather than soothe it, "and they knew what they were talking about!"[1]

This was the closest I ever heard the general come to complaining.

Meanwhile, Nan had shown me the greatest consideration. Nan Koch prided herself in being an independent, outspoken woman. One of her favorite anecdotes was about a time when, driving in Washington, D.C., she was somewhat slow in moving when a traffic light changed from red to green. The driver of the car behind her honked his horn impatiently. Instead of speeding away, she stopped, got out of her car and went back to the hurried driver and asked politely, "Did you want me for something?"

Nan did not hesitate to speak her mind. She did not like General Omar Bradley, and told a newspaper reporter he was "the skinniest, ugliest, corniest man you ever saw in your life. He might have been all right with some, but he wasn't my type of person."[2]

When we first learned of General Koch's illness, I was afraid Nan might be overly demanding of the doctors and expect more than the medical establishment was able to provide. My worry proved to be misplaced. She showed great patience and understanding, along with her concern that he get the best care available.

Although the general's health was foremost in both our minds, we recognized that he wanted work on the book to continue. We did our best to keep the project moving. She let me know that if I needed anything from his files she'd be glad to try and track it down—"But don't expect me to find it very fast!"

When the general came home, his first question to me was, "How's the old opus coming?" His interest in getting our book finished and into print had never waned. But as I looked at him lying in bed, pale and drawn and appearing to be very tired, my spirits plummeted.

I worried he might never be active again. and how could I proceed without him? I was reminded once more that this was his book, he was the authority, and my contribution was merely to put his thoughts and words into a form that might merit publication.

The general was severely weakened and bedfast for a time and I was reluctant to burden him with work on *G–2: Intelligence for Patton*. But at the same time I felt an overwhelming obligation to move ahead as rapidly as I could. The book must be finished before his life drained away and

I had no clue as to how long this might be.

I've often thought that if Oscar Koch had been in a field such as medical research, his name today would be associated with great discoveries. His thoroughness, his penchant for hard work, his attention to detail, his persistence and determination, all these qualities surely would have led him to solve at least some of the darkest mysteries of any disease he chose to study.

These same qualities were among the traits that made him an outstanding intelligence officer. His use of ULTRA intelligence is a case in point. By virtually all accounts, his use of ULTRA set a high standard.

Early in the war, the British had broken the code of the Germans' encrypting apparatus known as the Enigma machine. This meant that the secret German messages sent by wireless radio using the Enigma code, once intercepted, could be deciphered by the British. This information, classified Top Secret ULTRA, was channeled down to the level of army command through a network created specifically for this purpose. When the United States entered the war, ULTRA intelligence also was shared with the Americans.[3]

Somewhat like DNA in criminal trials today, ULTRA intelligence came to be overly relied on. Many intelligence officers began to relax their emphases on tried and true methods of collecting information. It was easier to depend on intercepted radio messages, which might give things like the precise strengths of German units, or their locations, or even reports on troop morale. Intercepted orders to move or attack, passed along by the British special liaison units, readily played into quick tactical decisions by American commanders.

In short, wrote U.S. Air Force Major Bradford J. Shwedo,

these officers "were so infatuated by this source that they disregarded their duties as G–2s and became full-time ULTRA officers." But not Oscar Koch. "When it came to ULTRA," Shwedo wrote, "Koch showed a maturity that other G–2s did not always possess."[4]

Oscar Koch's use of ULTRA intelligence germinated in North Africa, took root in Sicily, and blossomed into a superb model of effective application in the coming campaign across Europe. His method of incorporating ULTRA intelligence into his comprehensive system of using all available sources was a major reason for his success.

Patton's emphasis on rapid offensive movement placed a premium on timely information, which was the essence of ULTRA. In a contemporary report, Major Warrick Wallace, assistant head of the British ULTRA team attached to Third Army, wrote that, "An army has never moved as fast and as far as the Third Army in its drive across France, and ULTRA was invaluable every mile of the way."[5]

In the unique setup of the ULTRA system, intercepted German messages were processed and relayed to special British liaison units attached to a commander's headquarters and passed on to G–2. This meant the British retained control and the liaison units were a critical link without which ULTRA information never would have arrived in American hands.

Major Melvin Helfers, the ULTRA detachment commander, described British troops in earlier liaison units in Patton's commands as little short of slovenly. "They wore their uniforms as they pleased. They came and went as they pleased. They kept their quarters as they pleased...." Clearly, they did not manifest the level of discipline required by Patton, who had no control over them and had no great love for the British in general.

Above: Col. Oscar Koch poses before a plaque honoring his old commander, Gen. George S. Patton Jr., following dedication of Patton Hall at Fort Riley, Kansas, in 1946. Gen. Koch was head of the Army's intelligence school at Fort Riley.

Below: Gen. Koch as 25th Infantry Division commander in Korea.

Above: Gen. Oscar Koch as commander of the 25th Infantry Division in Korea. Below left: Gen. Oscar Koch at the time of his retirement from the U.S. Army in 1954 at age 57. Below right: Gen. Oscar Koch and Nan, early 1950s.

Photo courtesy Dr. Murray Zimmerman

Major Helfers, on the other hand, was a graduate of The Citadel and understood very well that "sharp uniforms, spotless quarters, and being a member of the greater team" were factors essential to success in Patton's headquarters.[6]

Under these circumstances, Koch basically kept the ULTRA detachment hidden from Patton in Africa and Sicily. Information from the British intercepts, by way of the liaison unit staff, came only to him. He evaluated it and incorporated it into his G–2 reports to Patton.

This changed soon after Third Army went into action on the continent. Major Helfers received ULTRA information on a potential German attack that he believed should be delivered to Patton immediately. He woke Koch, who agreed with his assessment. Koch took him directly to Patton. He recalled later that, after the briefing, Patton asked his G–2 why Helfers's presence in Third Army headquarters hadn't been revealed to him before. Koch, he said, told Patton that:

> since they had had such a bad experience with the British intelligence and signal troops attached to them in Africa and Sicily, he felt it best not to mention their presence and mission to him. General Patton took that as a reasonable and satisfactory explanation. General Patton then told me that beginning the next morning at seven I was to come to his trailer and present a short briefing on ULTRA intelligence and on what had come in over the past twenty–four

hours. I was to be prepared to do that until further notice.[7]

Direct ULTRA intelligence briefings continued to be part of Patton's daily routine from this point on, until the end of the war. According to a report by Major Wallace, Helfer's assistant, the ULTRA briefings followed the formal morning briefing in the war room. The latter session was attended by some forty staff officers, and at its conclusion the chief of staff would excuse all except Patton and seven of his senior staff members for a "special briefing." Then either Major Helfers or Major Wallace would spread the ULTRA map over the regular war room map and brief the commander and others in the exclusive gathering on the enemy situation as reflected by the intercepted and translated German radio messages.[8]

General Koch explained that the formal war room briefing itself was preceded by an informal gathering which led off the day. Patton and his key advisors would meet somewhere outside the war room for a session that typically lasted no more than fifteen minutes. Koch presented his intelligence report at this meeting, using a portable map or blackboard posted specifically for the purpose. His report would be followed by a period of discussion. Patton likely would ask a penetrating question—"What would happen if ...?" Staff members would respond, one question leading to another until everyone was on the same page.

"Most important," the general said, "we all knew what was on his [Patton's] mind, what he was going to do next."

Koch continued to use the ULTRA intercepts as supporting intelligence, not to be taken wholly at face value unless they fit the picture developed by "regular" sources.

Although ULTRA gave several dramatic warnings of German counterattacks, in the Third Army G–2 Section it mainly acted as a guide and critic to the mass of information from other sources.[9] This fit well with Koch's concept of all-source intelligence.

In the months ahead, this painstaking collection and analyses of information from every available source would lead Koch to provide his commander with an accurate and thorough reading of the enemy's capabilities. And he surely must have found a great deal of satisfaction in knowing that Patton had confidence in his work.

Chapter 14

THE ALLIES' CONQUEST of Sicily took only 38 days, and Oscar Koch's outstanding performance as U.S. Seventh Army G–2 was a major reason for the quick success. His best work, though, was still to come. Patton was relieved of command of the Seventh Army at the beginning of the new year, 1944, and his whereabouts kept secret. Koch was relieved of Seventh Army duty near the end of February on orders assigning him to a new but still secret command in England.

After a flight to Prestwick, Scotland, by way of Casablanca, he was driven to Knutsford, England. He found Patton just outside town in ancient Peover Hall, which soon would be the headquarters of the U.S. Third Army. Planning would begin immediately for a campaign against German forces occupying France. The invasion of the European continent was in the works.

We sat at the crowded dining room table in the Koch home and the general, looking more fit than he had for weeks, told me about his move from Sicily to England and his satisfaction at being reunited with Patton. He paused occasionally to sip the hot coffee that sat before him in a fine china cup. I had a sense that he was particularly happy this day, not just because he felt better physically, but because we were back at work on the book.

My efforts on sections of the manuscript I could draft from written records had not stopped, but this was our first face-to-face session in some time. He was eager to talk and I scribbled notes as fast as I could, hardly finding time for my own coffee.

When he described the flight to England as "uncomfortable but exciting," I wanted to know more. Was it exciting because it was dangerous? Uncomfortable because of the weather?

The general smiled. "I didn't say that very well," he said, his voice still weak. "The flight was exciting because I was on my way to rejoin Patton. It was uncomfortable for all kinds of reasons. Weather, yes, and I guess it might have been dangerous, too, but I never thought much about that. It was just a hard trip because a C–47 can bounce you around pretty good, and in the dark I could never tell where we were. It felt to me like we were just flying in circles."

"You probably were flying in circles," I said.

"Could have been. Pilots didn't like to fly straight lines from one point to another."

"For security?"

"Yes. Flying a straight line made it too easy to be seen and caught up with by German fighter planes."

I offered the proposition that Oscar Koch was hot cargo. No one wanted to be responsible for Patton's favorite G–2 getting shot down and lost. He wouldn't have any part of that argument. I believe it pleased him, though. And from my perspective, it was absolutely true. Patton was getting ready for an operation that was huge in scope. Koch was his tried and true intelligence officer, one he trusted implicitly, and losing him would be the last thing Patton wanted.

Patton had been given command of Third Army, which was still in the States at this time, and had begun a systematic effort to pull together a premier headquarters staff of handpicked officers. In addition to Oscar Koch, these would include a number of others from his former Seventh Army staff in Sicily. They would be integrated with some from the existing Third Army headquarters when that force arrived from Fort Sam Houston, Texas, in late March. In the meantime, Patton wanted those who already had joined him in England to get busy.

"I had hoped the invasion plans wouldn't change once we got a good start, like they did for Sicily," General Koch said. "But we were not that lucky. The plans for Third Army's role didn't change, but higher headquarters decided we needed to put together five alternate plans for landing on the French coast, just in case."

"Just in case? Just in case of what?" I asked.

"Just in case the main offensive by First Army failed."

"Backup plans."

"Yes, backup plans. For landings at five different sites."

What the general was talking about, of course, was the cross–channel invasion of occupied France. The critically important undertaking would be called Operation Overlord. Regardless of which plan ultimately was put to use, it would be a joint operation that involved seaborne assaults coordinated with air and naval forces.

Shortly after Koch arrived in England, Patton identified Third Army's target, deep in France.

"I was summoned to the Old Man's office and when I

went in he was leaning over a table alongside an inner wall, his back to me. I waited until he straightened up slightly, and asked me to join him. He kept the index finger of his right hand on a Michelin road map of western Europe and said, 'Koch, I want all of your planning directed to here.' His finger was on Metz. That was all I needed to know."

In spite of unexpected events during the eventual campaign across France, Patton never deviated from this goal. Third Army's successful assault on Metz 11 months later would prove the success of the commander's tenacious planning—planning based on Koch's heady intelligence projections.

Patton's edict, which turned out to be the only intelligence directive he would issue personally throughout the war, meant that Koch and his G–2 Section staff must look for anything that might affect the Third Army mission, from the coast of France, across Brittany, all the way to Metz. The starting point would be the basic elements of intelligence—the enemy, the terrain, and the weather.

I said something facetious, like, "Simple, right?"

General Koch knew I was joking. But he went on to explain what was involved, based on his own incomparable experience. Patton, commanding a cross–channel invasion force, would be concerned first of all with the Germans' naval defenses. The following paragraph in *G–2: Intelligence for Patton*, based to a great extent on notes I took at this time, most certainly would have been appropriate for the text book on intelligence in combat the general was writing before we decided to broaden the scope of the new work:

> For example, asking about the extent and nature of enemy naval defenses and activities in

the area led to specific additional questions concerning such items as the number, type, and condition of captured or salvaged craft, the number, type, location, and effectiveness of German naval units, and the status of training and the morale of the crews. What types, to what extent, and what means of off-shore gunfire protection were employed? How were underwater defenses controlled? What were the types of obstructions used under water, in ports, on the beaches—wire, mines? And what was the number, type, direction, and time of enemy movements by sea and close–in inland waterways?[1]

In addition to the intelligence information crucial for Patton's coming operations, the Third Army G–2 Section got requests directly from higher headquarters for information specific to their needs. All in all, it was a tall order.

About a month later, Koch learned that Third Army would be a follow–up force and would commence action thirty days after the U.S. First Army's landings on the continent. When that might happen was still a closely guarded secret. Koch, never one to leave something to chance, nevertheless assumed he had at least ninety days to complete the essential intelligence collection and analyses before Patton's troops saw action.

"We had the complete intelligence plan for Patton's drive into France done by late April," General Koch told me. "It wasn't a static plan, of course. We had to keep it updated until the day came when it would be put to use."

"Meaning lots of changes?" I asked.

"Lots of changes. We had to operate every day as if we might see action tomorrow."

The story of how Patton was used as the major element of a classic deception is well known. To the Germans, Patton was the most feared and respected Allied general. Hitler believed the Normandy invasion was a feint and was confident Patton would lead the main Allied assault, launched in the Pas–de–Calais area.

Allied headquarters went to great lengths to perpetuate the fraud, maintaining fictitious radio traffic that would lead the enemy to believe a U.S. Army Group was standing at the ready in southeast England, opposite the Pas–de–Calais. Hitler held his Fifteenth Army with its eighteen divisions in position to defend against that phantom American force, waiting for the Patton invasion that never came. Tough as the D–Day landings were, they could have been far worse had the deception not worked. Third Army joined the battle on August 1.

Patton's forces poured into France, with two corps moving rapidly down the west coast of the Cherbourg Peninsula. With only two main coastal highways, they faced a bottleneck where the roads converged at the Avranches Bridge. There was only a single highway for a distance of about five miles.

To the east, elements of the U.S. First Army were heavily engaged with the German forces. The enemy was meeting them with a determined defense.

Koch reported in his Third Army G–2 Periodic Report issued August 3 that there was serious potential for the Germans to launch a counterattack. If they did, it would be a grave challenge on Third Army's left flank. He worried that he hadn't been able to pin down the location of four Panzer divisions, and "Such a concentration would give the enemy a force capable of being used in a major counterattack aimed at driving a wedge to the Channel between our North and South forces."

Where were the four Panzer divisions? When and where would they be used?

General Koch had used this episode in an earlier manuscript as an example of intelligence at work on the battlefield and we kept a somewhat abbreviated version in the new book. He described how he was awakened at 2 a.m. by the G–2 Section duty officer, who had critical information. Higher headquarters had just heard from a "usually reliable source" that a German counterattack of major proportions against the First Army was imminent. The First Army had been informed.

We know now that the "usually reliable source" he cited was ULTRA. This was the occasion on which Koch first took Major Helfers of the British ULTRA detachment with him when he went to brief Patton. (And it also is the episode Ladislas Farago asked Koch about years later, mistakenly relating it to the beginning of the Battle of the Bulge.)

Patton's quick response effectively headed off the Germans' violent attack by four Panzer and one Panzer Grenadier divisions and created what would come to be known as the Falaise Gap, where the entire German Seventh Army was surrounded except for a narrow escape route. Patton's XV Corps was in a position to close the gap within a matter of hours. Higher headquarters, instead, ordered a halt to Patton's drive and assigned the task to Field Marshal Montgomery. It took Montgomery's British forces several days to close the gap and the Germans escaped.

"What the Germans could not do," wrote Robert S. Allen, "SHAEF did." Allen attributed SHAEF's poorly considered decision to high-level politics.[2]

This failure went even further when SHAEF denied American P–47 Thunderbolt fighter pilots permission to attack more than two thousand parked German vehicles because they were "outside" the American area of responsibility.

Montgomery's slowness let the Germans recover troops, tanks, and artillery that Patton's forces would have to fight all the way across France.[3]

There was nothing left for Patton to do, in Oscar Koch's words, except to pursue the fleeing enemy. Third Army troops advanced so rapidly on Paris that a planned air drop proved unnecessary. Paris fell to the Allies on August 25—less than four weeks after Third Army was activated on the continent.[4]

Whether he was talking about Patton's battle tactics, his personality, his reputation, or whatever, Koch enjoyed discussing his old commander. He quickly became more animated and his countenance brightened. His references usually began with "General Patton," then turned to "Patton," and in a short time, "the Old Man." Each of them was spoken with immense respect, or even outright reverence.

Although I would have liked to have a complete transcript of every word General Koch had to say—potential material for our book—I was reluctant to use a tape recorder. That might have added too much formality and dampened our discussions. He talked freely, and he knew I was making notes, but pen and paper are much less menacing than a microphone in someone's face. Sometimes I got caught up in the conversation and failed to keep notes as complete and detailed as I intended. That left me to try and fill in the gaps later. I always was able to remember the substance of what he said, if not his exact words.

The general's "good days" were coming more often now. I could see that he still tired quickly, though, and there were instances when I made excuses for having to leave early just to make sure and not wear him down.

This meant I couldn't spend nearly as much time with him as I wished, combing through his material, going over first drafts of copy for the book. He was sensitive to the problems this posed for me and insisted that I should keep writing full-speed. He would get around to reading everything "in good time." Fortunately, he already had written a good deal about the early Third Army campaign. Some of his writings were rich in detail and they all related to things I wanted to know more about.

As I considered his prognosis, I became even more grateful for his time. I wanted to take full advantage of his availability, because the future was uncertain. But for now he was regaining his strength and vitality. I could almost forget the cancer that we assumed still lurked somewhere beneath the surface, just waiting its time. Almost.

Chapter 15

MY WORK WITH General Koch was a learning experience on many fronts, not the least of which was geography. He knew the map of Europe like the back of his hand. I usually had only a dim idea of where things were. I'm sure my problem was shared by many Americans then as now and exacerbated by the easy misperception of European distances. We live in a nation that is some three thousand miles wide and too easily forget that France's borders are less than six hundred miles apart, east to west.

I understood Third Army's drive better once I had figured out some of the statistics. For example, from Avranches to Metz, the objective Patton had set when Koch first joined him in England, is about seven hundred kilometers, or less than 450 miles. I had an easy comparison. My home state, Illinois, is almost four hundred miles long, north to south.

Distance can be misleading, of course. Patton's forces had to fight their way across France against heavy German resistance. Mile by mile, and sometimes, even, virtually yard by yard.

The liberation of Paris in August 1944 paid big dividends for his Third Army G–2 Section. Metz was a heavily fortified city. With Paris safe in Allied hands, Koch and his staff had access to French files and archives that contained exhaustive studies of the fortifications that lay ahead as well as an array of detailed maps of the area.[1]

I said, "It's a good thing all this wasn't destroyed by the Germans, or by a battle to free the city."

"No doubt about it. Having Paris spared from destruction was one of the most fortunate things that happened in World War II. If you'd seen what little was left of so many of the beautiful old German cities you'd really understand how lucky that was."[2]

"And French cities, too."

"Yes," the general said, "some of the French cities, and the little towns too, took heavy damage in battles we had to fight to liberate them from the Germans. Many of the German cities could have been spared if Hitler hadn't been so fanatical. His generals knew the war was lost and would have surrendered to save their country but Hitler didn't care."

When Third Army troops driving eastward met up with the U.S. Seventh Army moving up from the south, they crushed the German 16th Division between them. More than 19 thousand German troops were taken prisoner. Many of the POWs were put to work in a supply depot near Reims repairing and refurbishing war materiel badly needed at the front. A large share of the POW administration, both in the prison and in the shops, was placed in the hands of captured German officers. Food, medical supplies, fuel, and communications wire seized from the Germans were allocated to Third Army, and in late August more than a thousand planes were used to deliver rations and fuel to Patton's forward elements.[3]

"We had the feeling that nothing could stop us," General Koch said. "And it wasn't the Germans that did."

In spite of the Herculean efforts of airborne freight pilots and the famed Red Ball Express with its six thousand vehicles from 67 Transportation Corps truck companies, necessary supplies could not be delivered fast enough to keep pace with the rapidly advancing Third Army. Patton's forces had dashed across France, liberated the Brittany Peninsula, and seized bridgeheads east of the Meuse River by the end of August. It now occupied a position Allied planners had not expected to reach before April, just over a hundred miles from the Rhine.[4]

The German troops fell back in orderly fashion and kept up a strong fight. Patton still faced the challenge of breaching the Siegfried Line, but information gained from ULTRA as well as from POWs indicated these vaunted defensive positions were not prepared for attack. In some instances, the German field commanders did not know where the line's key emplacements were located.[5]

Third Army was poised for a final dash into the German heartland. But once again, higher headquarters—his own superiors—slowed Patton down.

"We weren't getting enough gasoline to run on at best, then Eisenhower decided to send most of what there was up North to Montgomery and Hodges," General Koch said matter-of-factly.

Montgomery was about to launch the Operation Market Garden offensive in Holland. He would be supported by the U.S First Army, now under the command of General Courtney Hodges. Eisenhower elected to give priority to Montgomery's offensive—a decision that was to become even more controversial when Market Garden proved a dismal failure.

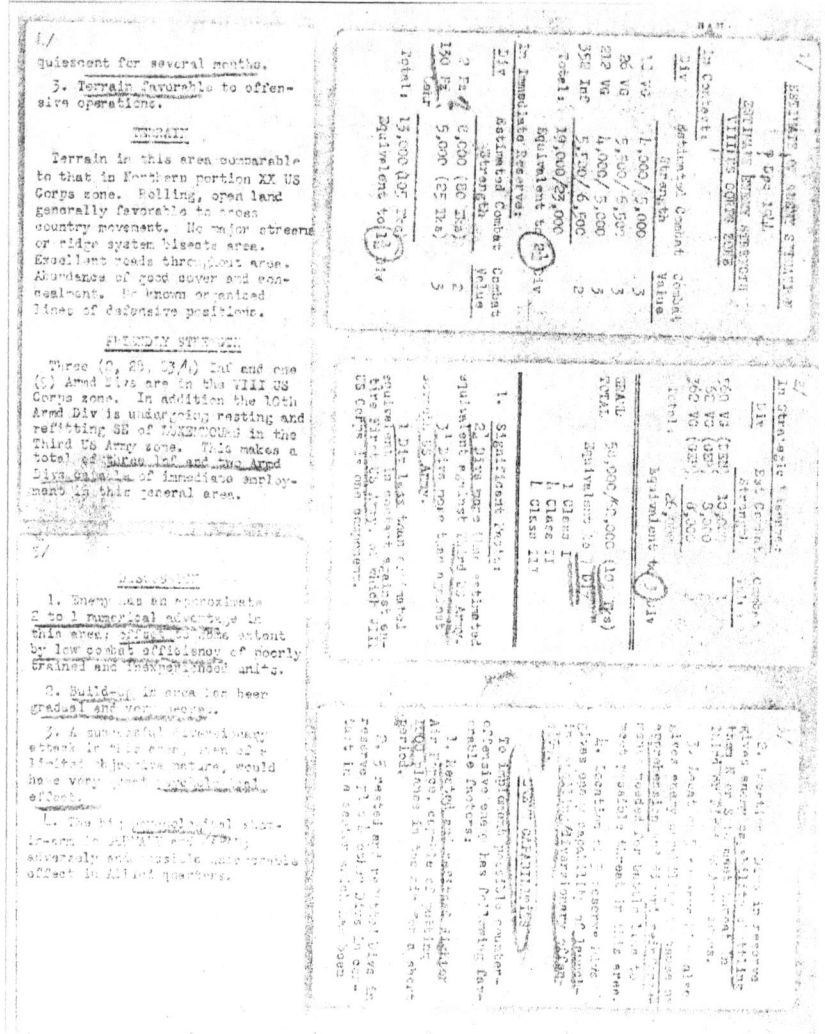

Col. Oscar Koch typed his notes on index cards in preparation for Patton's early morning intelligence briefings. This faded copy from Koch's files shows cards used for his report on December 9, 1944, when he outlined the full extent of the German buildup in the Ardennes a week before opening shots were fired in the Battle of the Bulge. Higher headquarters continued to ignore Koch's warnings, but Patton directed his staff to begin making plans for an enemy counteroffensive.

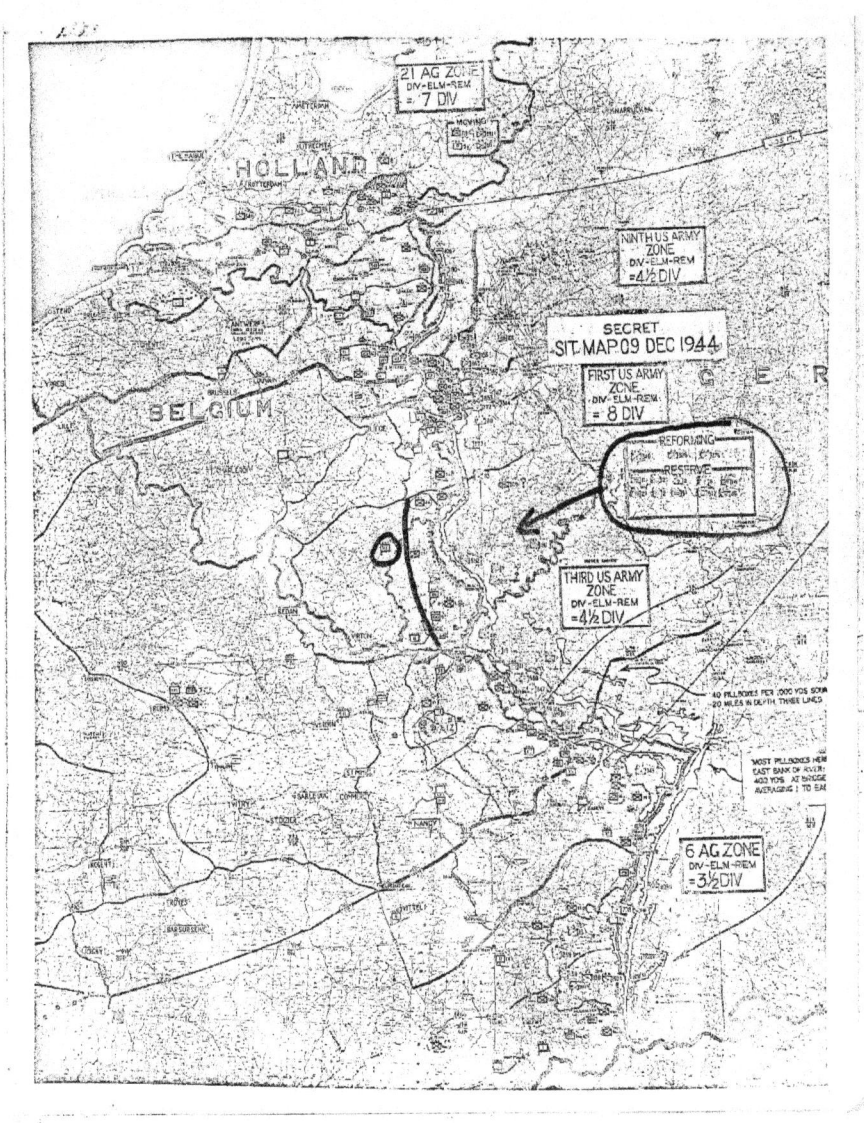

Faded photocopy of the U.S. Third Army G–2 Section situation map on December 9, 1944, from Gen. Oscar Koch's files, marked to highlight the strength of the German buildup preceding the Battle of the Bulge.

Patton protested, but his dissent fell on deaf ears. Barely half of the 14 thousand tons of supplies a day needed to stock both Third and First armies was being delivered, and under Eisenhower's new rules two-thirds of this went to First Army.[6] Patton's forces were stopped in their tracks and virtually shut down. As one immediate result, a Third Army reconnaissance outfit, the 15th Cavalry Squadron, which had entered Metz on September 1 with little resistance, had to withdraw when lack of fuel prevented supporting units from joining it to help hold the city. Predictably, this would have disastrous repercussions later.[7]

While Third Army remained at the bottom of the supply priority list for weeks to come, Patton's troops, with no gasoline for their tanks, could do nothing in terms of forward movement. They were left to attack areas where limited offensive action, principally by infantry units without the support of armor, might reduce some of the stronger pockets of resistance or deny the Germans an opportunity to constitute new ones.

"The best we could hope for was to get rid of whatever defenses we could and try to make forward movement easier once we got the go-ahead from higher headquarters," Koch explained. "This wasn't what we wanted, but we had no choice."

Patton also used the enforced down time to get his soldiers ready for the coming winter. This included ordering winter clothing. Colonel Walter J. Muller, Third Army quartermaster, requisitioned winter dress for the troops. Patton protested vehemently when he learned that supplies of rubber overshoes were issued on the basis of one pair for every four men, and a "more reasonable" issue was established.[8]

Koch took advantage of the lull in action to have his G-2 Section do a comprehensive study of the Metz fortifications. These used the Maginot Line, originally built by the French to protect against the Germans. The old installations would have been modernized, Koch warned his commander, and it would take the full force of modern weaponry, mobility, and air power to overcome the enemy defensive positions.

He was right. With Patton's troops at a near standstill, the Germans made the best of their opportunity to regroup and rebuild the historic Metz fortifications.

When SHAEF finally lifted the restraining order against Third Army in November, Metz and its twenty-plus surrounding forts once again became Third Army's immediate primary objective. This time it took ten days of vicious fighting, but Patton's forces took Metz by direct assault for the first time since 451 A.D.[9]

The victory was costly. As Robert S. Allen put it, "What could have been accomplished quickly and easily in September, when Third Army was in full smashing stride and the Germans on the run, now had to be bought with blood."[10]

Meanwhile, Koch reported that the Germans had some forty thousand slave laborers reinforcing installations of their next major point of defense, the Siegfried Line. This meant another series of strong fixed defensive positions ahead for Third Army troops. It looked as if Eisenhower's decision to favor Montgomery over Patton would prove even more costly in the weeks to come.

I might have expected General Koch to express some outrage over the injustice higher headquarters did to Third Army in virtually shutting off its fuel supplies. But outrage was an emotion Oscar Koch rarely if ever showed. He had written, in reference to the shutdown, "Problems of supply had grown acute and the cessation of major offensive operations would give logistics an opportunity to catch up."

Others have been a great deal less generous. Major Bradford Shwedo, echoing Robert S. Allen's analysis of the earlier failure to let Patton close the Falaise Gap, wrote that when SHAEF cut off Patton's supplies and stopped Third Army in its tracks "the Allies were finally able to accomplish something the Germans were never able to do."

Patton himself wrote, "No one realizes the terrible value of the 'unforgiving minute' except me. Some way I will get on yet."[11]

Perhaps the strongest indictment of Eisenhower's decision came from those who benefited most, the enemy. German Field Marshal Karl Gerd Von Rundstedt wrote that "at this time the Allies could have broken through the West Wall [the Siegfried Line] whenever they wished.... But to our great surprise the operations of the Allies came to a full stop."[12]

General Richard Schimpf, commander of an elite German paratroop division, wrote after he was captured by the Allies that the Germans "confidently relied on Allied hesitancy to exploit successes to give us the time to withdraw and regroup in order to slow up the next thrust." He explained:

> There is no question that if your Third Army had not been halted before Metz in September, it could have penetrated the Siegfried Line very quickly and been on the Rhine in a short time. At that time we were powerless to

cope with the situation in that portion of the Front. But when your Third Army was halted, we obtained the time to regroup and we used that opportunity to the utmost.[13]

Patton would complain later that Eisenhower "kept talking about the future great battle of Germany, while we assured him that the Germans have nothing left to fight with if we push on right now. If we wait there will be a future great battle of Germany."[14]

With Metz occupied and Third Army again on the drive, Koch turned the intelligence section's sights on Frankfurt-on-Main, the objective—by way of Kaiserslautern and the Palatinate—of a new offensive set to kick off in mid-December. Ahead lay the demoralizing defenses of the Siegfried Line, now significantly strengthened.

Then the kind of break intelligence officers frequently dream about, but seldom experience.

Koch learned through a routine report from higher headquarters that a captured German officer had detailed information about the Siegfried Line and was willing to share it. The circumstances were like those the general had talked about in our discussion of the destruction of Paris. In this case, the prisoner believed Germany had lost the war, but feared Hitler would continue to fight. He wanted to do whatever he could to help shorten the conflict and spare his homeland further destruction.

Koch had the prisoner brought to Third Army headquarters for questioning.

The initial report was accurate: the prisoner was willing to share what he knew. And what he knew proved to be remarkable. Years before, he had been involved in fortifying the enemy defenses that lay directly in the path of Third Army. His memory was excellent. He could point out on air reconnaissance photos the field fortifications and knew the details of their construction. He even knew the types of weapons they had been designed to accommodate and their fields of fire. He also was able to show the sites of some fortifications that did not show on the aerial photographs.[15]

Koch's G–2 Section checked and verified the information the German prisoner had provided. Everything was printed on large-scale maps and made ready for distribution to all units scheduled for action in the major offensive Patton's troops were about to launch.

"But we never used it," General Koch said.

"You never used it? Why?"

"We were a little sidetracked. Remember, we're talking about December 1944."

"The Battle of the Bulge!" I said.

"Yes."

The information Oscar Koch gained from the captured German officer was not wasted, however. It was transferred to the U.S. Seventh Army. And given what was to come, as we wrote in *G–2: Intelligence for Patton*, Seventh Army used the information "as originally intended—to its fullest advantage."

As long as it was needed, good intelligence would never lose its value. And Oscar Koch provided good intelligence.

Chapter 16

As the weeks and months rushed by, General Koch's cancer appeared to be in remission. He was up and about on a full-time schedule, he had regained his appetite, his color was good, and in every visible characteristic it seemed as if he had returned to a state of good health. He'd never lost his sense of humor nor his ability to see the best in the world around him.

Deep down, I knew this wouldn't last. I pretended it would. I tried to keep from thinking about that day I knew would come, that day when this deceitful mask would fall away and the true vision of the general's final days would be revealed.

Nan asked if I could come by the Koch house on a Saturday morning and observe a formal flag-raising ceremony the general was directing for a local troop of Boy Scouts. She had mentioned several times that he always liked working with young people. I got there some twenty minutes early, and I could see that he genuinely looked forward to the event. It was clear to me he found pleasure both in working with the scouts and in teaching proper handling of the flag. The event itself actually was rather impressive.

I had been busy on various sections of the *G–2* manuscript, bringing copy to the general only a few pages at a time, but now he said I could work him harder. He reviewed the pages quickly, made such notes as he felt would be helpful, and had them waiting well before I was ready for them again.

Most of what I was working with came directly from our collaborative sessions or a section of his earlier manuscript, so there seldom was much rewriting for me to do. It was especially good to have the general more readily available again. We were about to go to work on what both of us considered the most important topic in the book: Koch's Third Army intelligence work in the weeks and months before the Battle of the Bulge.

"Of all the time I spent with Patton, this may have been the most critical," the general said. "If he hadn't trusted my work, the Bulge could have turned out a lot worse than it did."

"But it was bad enough," I said.

I had done my homework. The German offensive that came to be called the Battle of the Bulge extracted a terrible price from the Allied troops.

"It never needed to happen."

"I know. I mean, I'm beginning to find out. The record has to be set straight."

The general appeared to hesitate, as if considering what he should say next. This was not something I was accustomed to. He always was miles ahead of me, always knew exactly how to lay out the story he wanted to tell.

"I want to show you something," he said.

On the table nearby were stacks of material he had been working with, penciling notes on his customary yellow tablet. This was his usual method and it was the yellow tablet pages I normally worked from. His handwriting was easy to read and his notes were organized in a logical fashion, so I seldom had a problem with his material. I often wrote several pages of manuscript text from a couple of pages of his notes.

He moved a yellow tablet aside. I could see he had been reading copies of some of the original Third Army intelligence reports. I felt a wave of gratification that he had saved them, that we had them at hand for reference. These were the real thing, holding precisely the same information Patton and the other commanders had in December 1944.

But the general pushed the intelligence reports aside, too. Hidden beneath them were some standard white index cards.

"These are the actual notes I used to brief Patton," he said, sliding the cards across the table.

What I saw before me were five two–inch by three–inch index cards that held the typewritten notes Oscar Koch had used in an historic intelligence briefing in Third Army headquarters in Nancy, France, on December 9, 1944. There was enough detail to provide ample warning of the German buildup in the Ardennes area a full week before the enemy offensive began. Koch had been watching the enemy buildup since October.

"I always made it a practice to keep briefing materials so I could check their accuracy against future events," the general said. "Sometimes they were very useful."

Although I still had much to learn about the situation Third Army and other Allied forces faced that day, I already knew enough to understand the significance of what lay before me. Here was indisputable evidence that most of those writing about the Bulge for the last quarter-century had been wrong.

Contemporary reporting and post-war histories, with very few exceptions, claimed there had been an Allied intelligence failure preceding the German offensive, giving Hitler's forces the advantage of complete surprise. American and British news correspondents at the front had routinely filed stories reporting the pre-Bulge complacency of Allied intelligence officers, supposedly bred by exaggerated confidence in the superior power of their own forces. Intelligence had completely failed to detect the enemy movements, they said, and press criticism of the intelligence operations became vehement.

At least one editor, though, took issue with this position. The *San Antonio Express*, in a late December editorial, claimed intelligence had been made a scapegoat because few of those making that charge "have dared blame General Eisenhower and his staff." It went on to note that newspapers for some time had been publishing stories that Von Rundstedt was holding his top grade armored divisions in reserve, presumably to defend against the inevitable Allied advance. This obviously undermined the SHAEF claim there was no strength in the German buildup.[1]

Eisenhower, in his 1948 book, *Crusade in Europe*, writes that through late November and early December he and Bradley had been convinced that the risk of a German breakout in the Ardennes didn't merit giving up the Allied offensive. He adds that Bradley "further estimated that if the enemy should deliver a *surprise* [emphasis added] attack in the Ardennes he would have great difficulty in supply if he tried to advance as far as the line of the Meuse."[2]

Winston Churchill, in his prodigious six-volume history, *The Second World War*, says Hitler's stalwart Sixth Panzer Army, which would spearhead the German counterattack, "vanished for a while from the ken of our Intelligence, and bad flying weather hindered our efforts to trace it. Eisenhower suspected that something was afoot, though its scope and violence came as a surprise."[3]

In his 1966 book, *Battles Lost and Won*, Hanson W. Baldwin writes that the "Nazi attack in the Ardennes was a shattering surprise.... the enemy's offensive strength was discounted." To his credit, Baldwin, the military editor and analyst of *The New York Times*, does acknowledge that "The U.S. Third Army G-2, Colonel Oscar Koch, came closer to the mark." Ironically, dust-jacket copy of Baldwin's book boasts that it "combines most effectively the emotional and dramatic immediacy of tremendous events with the knowledge and wisdom of afteryears."[4]

Even the Army's official history of the Battle of the Bulge, published in 1965, contends that:

> The prelude to the Ardennes counteroffensive on 16 December can only be reckoned as a gross failure by Allied ground and air intelligence. One of the greatest skills in the practice of the military art is the avoidance of the natural tendency to overrate or underestimate the enemy. Here, the enemy capability for reacting other than to direct Allied pressure had been sadly underestimated.[5]

The British military historian B. Liddell Hart comes closer to stating the situation accurately. He, too, says the attack came as a "shock" to the Allies, but rightly ascribes the problem to the fact that "some of their highest commanders had been confidently saying that the Germans would never be capable of another offensive."[6]

Among the few who had the story straight, in addition to Robert S. Allen and Ladislas Farago, are the editors of *Army Times*. Their book, *Warrior: the Story of General George S. Patton Jr.,* published in 1967, tells the story accurately. They write that "Much of Patton's success was due to brilliant staff work" and contend that Patton's staff was "at maximum effectiveness in early December when Colonel Koch first warned ... of the ominous buildup of German combat power opposite First Army in the Eiffel region." And further, they write that while higher headquarters "scoffed" at Koch's reports, Patton and the Third Army staff "immediately got to work on a new problem which was not yet theirs to solve."[7]

On balance, though, Oscar Koch had received precious little credit for his remarkable pre-Bulge performance. I suggested that he deserved a great deal more. He gently disagreed, and it was obvious he was firm in his conviction.

"Credit to me is not important," he said. "What's important is the example. It wasn't intelligence that failed. It was what the top command did with the intelligence we had."

"Top command? You're talking about Eisenhower and Bradley, right?"

"I don't like to point fingers. General Patton personally called Bradley's attention to the situation in a phone call right after our briefing. But you have to understand the terrible pressures we all were under at that time. There just weren't enough hours in a day for us to get done all the things that needed to be done. This was true for staff members at Third Army level; it must have been even worse for an Army Group commander [Bradley].

"Some G–2's complained that there were times when they reported things to their commander and he sat there and looked them in the eye without ever really hearing what they said, and would ask later, 'Why didn't you tell me that?' He had too many things on his mind. Those were the kind of days we were living in. Bradley logically would have passed the information on to Sibert [General Edwin L. Sibert, Bradley's G–2] and then forgot about it until it came up again—which it apparently never did."

I wanted to use his explanation in the book. I'd been taking notes as he talked.

General Koch had other ideas.

"No, don't use it," he said. "I don't want to put the monkey on Bradley's back."

Oscar Koch apparently held a positive opinion of General Omar Bradley. In writing about action in North Africa, he noted that Bradley followed Patton as II Corps commander and added that Bradley was "later to become Patton's commander and to prove one of the Allies' most able field generals."[8]

I've always tended to be much more critical of Bradley, who eventually became the first chairman of the Joint Chiefs of Staff. In summing up the failures preceding the Battle of the Bulge in a chapter in *G–2: Intelligence for Patton*, I wanted to cite the lack of awareness shown by Eisenhower and Bradley. The general asked that it be changed to "higher headquarters."

"Besides," he said, "we don't have to blame anybody. We just need to give an accurate account of the information

Third Army intelligence had and what happened."

"Yes. And let the chips fall where they may?"

"Let the chips fall where they may."

Starting with the general's index cards and supported by the written Third Army intelligence reports from those bitter winter months of 1944, I was confident we had everything necessary to prove conclusively that Oscar Koch and his G–2 staff had pieced together and disseminated detailed evidence of the latent enemy preparations in the weeks and months preceding the Bulge, but their reports had fallen upon deaf ears and gone unheeded at higher headquarters. Patton, on the other hand, had confidence in their work and prepared accordingly.

And there was more. Not only did General Koch have his own contemporary intelligence materials, but he also had the results of his study of the German military archives. Item by item, he had compared the German records of what actually was happening at the time to his intelligence reports on what he and his G–2 staff thought was going on. His Third Army intelligence reports had been remarkably accurate.

Details of Koch's brilliant intelligence work in the period leading up to the Battle of the Bulge are presented in three of the most important chapters in *G–2: Intelligence for Patton*. Koch's pre–Bulge reports culminated in his G–2 Periodic Report No. 188, issued on December 16 and covering the period ending the previous midnight. Koch predicted that "The enemy is massing his Armor in positions of tactical reserve presumably for a large–scale counteroffensive...."

Before dawn that same day, Hitler's artillery opened fire in the Ardennes. Then came the panzers, spurting out of the snowy forests and searing through American lines. The Battle of the Bulge had begun. Before it was over, more than 77 thousand American soldiers would be dead, wounded, or captured as obscure Belgian villages such as St. Vith, Malmédy, and Bastogne were memorialized in blood. The Allied drive on Berlin ground to a halt, while Russian forces moving on Germany from the east gained an advantage which in time would have profound consequences on the division of conquered Europe.

Patton had acted immediately after Koch's December 9 briefing. Not waiting to see what Bradley and Eisenhower would do, he ordered his staff to begin making contingency plans for meeting a German assault. Third Army was prepared to forego its own offensive and do an abrupt right-angled turn to take on the enemy attack to the north. When Eisenhower finally called a meeting of the Allied leaders at Verdun on December 19, others were skeptical of Patton's ability to respond to the German offensive as rapidly as he said he could. They did not know Third Army already had plans in place and merely awaited orders to move.

Surprisingly, Eisenhower contends that at that December 19 SHAEF meeting at which he finally took action, "Patton at first did not seem to comprehend the strength of the German assault and spoke so lightly of the task assigned him that I felt it necessary to impress upon him the need of strength and cohesion in his own advance."[9]

Bradley, in his book, *A General's Life*, written in collaboration with Clay Blair and published in 1983, notes that Patton had made a diary entry concerning the German buildup in the Ardennes on November 24. In a gross understatement, he says Patton's intuition may have been based, in part, "on the excellent work of his G–2, Oscar Koch." He goes on to enumerate Koch's intelligence reports during this period, while defending his own lack of action.

"I, of course, did not see these Koch reports," Bradley continues. "Even if I had, they would not have unduly alarmed me." Why? Because although Eisenhower's G–2, General Kenneth Strong, had warned Bradley that the German 6th Panzer Army might launch a spoiling attack through the Ardennes against the VIII Corps, "at no time did anyone present me with unequivocal or convincing evidence that a massive German attack" was imminent.[10]

Over time, following publication of *G–2: Intelligence for Patton*, the true story eclipsed most of the existing misinformation. While historians have not been reluctant to take Oscar Koch's word on the matter—after all, wide dissemination of intelligence reports was standard operating procedure—the actual delivery of the most revealing Third Army intelligence reports to higher headquarters is verified in the book, *I was with Patton*, D. A. Lande's compilation of first-person accounts by those who served in the Patton commands in World War II, published in 2002.

In one of those accounts, Warrant Officer Fred Hose, who served in Oscar Koch's G–2 Section, tells how he personally hand-carried the critical Third Army intelligence reports to General Edwin Sibert at 12th Army Group headquarters in Luxembourg. Insofar as later reports claimed that "people who were supposed to be in the know weren't aware of the [German] troop build-up," Hose says flatly that this "was crap."[11]

The once prevailing thesis that Allied intelligence was asleep at the wheel in the days before the German offensive was replaced with the important question: Why did Eisenhower and Bradley fail to act, when the same intelligence available to Patton was available to them? This is an important question for future military commanders because SHAEF's inaction no doubt cost the lives of thousands of American soldiers.

There has been a great deal of speculation that over-reliance on ULTRA by the higher commanders' intelligence staffs was a major factor in their not recognizing the threat of a German counterattack. Because Hitler had imposed radio silence on the operation, ULTRA picked up very little of the information Oscar Koch gleaned from other sources, including heavy reliance on air reconnaissance and POW interrogations. He also gained crucial information from the OSS detachment at his command.

Group Captain Winterbotham, whose book first made public the existence of ULTRA, writes that the absence of German radio intercepts in the pre-Bulge period "probably was the most pertinent cause" of SHAEF's failure to accept intelligence reports such as Koch's. He says those in the higher headquarters perhaps had come to rely on ULTRA to such an extent "that when it gave no positive indication of the coming counterattack, all the other indications were not taken seriously enough."[12]

Bradley also substantiates this view. He writes that, "One major fault on our side was that our intelligence community had come to rely far too heavily on Ultra to the exclusion of other intelligence sources."[13]

Although General Koch never complained about the failure of historians to give him credit for his outstanding work in the weeks and months before the German breakout in the Ardennes, I think he was somewhat puzzled by it. He also felt that a thorough analysis of what happened was important.

He pointed out that both Eisenhower and Bradley were convinced that at this point the German Army was too weak to launch a major counteroffensive and Hitler would use whatever reserves he could muster for a last–stand defense in areas where the Allied forces were strongest. A SHAEF Intelligence Summary issued in late August claimed the end of the fighting in Europe was "within sight, almost within reach. The strength of the German Armies in the West has been shattered."[14]

"If Ike hadn't shut down Third Army and sent your gasoline to Montgomery, they might have been right," I said.

"That's another story."

The general ignored my sarcastic interruption.

We were talking about the failure of higher command in the lead–up to the Bulge. He was more inclined to hold the Eisenhower and Bradley intelligence staffs responsible, given that they were steeped in the philosophy of judging enemy capabilities and not intentions.

"The American intelligence procedure was to reserve consideration of enemy intentions for the commander," he said. "It was intelligence's responsibility to say what the enemy *could* do and let the commander gamble on what he *would* do. We had ample evidence that the Germans were capable of launching a strong counteroffensive, regardless of whether they intended to, and this should have been clearly spelled out to Eisenhower and Bradley."

"But you assume it wasn't," I said. I intentionally did not pose this as a question.

"No, it wasn't. I talked to Strong [Eisenhower's intelligence officer] in England after the war and he said, 'How did we miss the Bulge?' I said *we* didn't, and he said something to the effect that his job had been too demanding and that if he'd had time to visit lower headquarters personally maybe situations like we're discussing right now could have been avoided. I think he was exactly right."

"And Bradley?"

"What can I say? Sibert didn't have an intelligence background when he joined Bradley's staff. Maybe he didn't do his job as well as he should have."

General Sibert, an Arkansas native and West Point graduate, served during the pre–war period as a military attaché in Brazil. Before he became Bradley's G–2, he was commander of the 99th Infantry Division's artillery. He became an assistant operations director for the CIA after the war.

Eisenhower biographer Stephen Ambrose, in his book *Citizen Soldiers*, says Eisenhower and Bradley "told each other that an Ardennes attack would be a strategic mistake for the enemy." But, "Had they looked at the situation from Hitler's point of view, they would have come to a much different conclusion."[15]

But it was not in Oscar Koch's nature to place blame. Although I recall him getting a good laugh from a letter he received from Robert S. Allen in which Allen implied that Eisenhower was "deaf, dumb, and blind in the use of intelligence," he was reluctant to be overly critical, himself. But he took charges there had been an intelligence failure before the Bulge very seriously, and felt obligated to pass on what he'd learned so that future intelligence officers would not let the same mistakes be made.

"If Patton had lived," he said, "some of the misconceptions about American intelligence performance in the pre–Bulge days would have been dispelled long before now. But I'm afraid time is running out for the first–hand story to be told."

I vowed to myself, silently, that this would be my principal goal as we went forward on *G–2: Intelligence for Patton*. General Koch was not looking for glory. He wanted the record to be set straight, though, and if we did that the glory would follow in due course. He was my hero. I would sing his song.

Chapter 17

As I sat at the dining room table in the Koch house one cold November afternoon, taking notes on material from some of the general's files, he suddenly stood and, without saying anything, left the room. He was gone for a few minutes and when he returned he said casually, "Here, I have something for you." He handed me a small card with a Christmas greeting from Patton printed on one side and a prayer for good weather printed on the other.

I value this little card today not only because of its inherent quality as an historic artifact of World War II, but also because it was a gift from the general. I know the prayer card was important to him, too, and always assumed he didn't give me his last one. There was no way to know.

A quarter–million of these cards were distributed to the soldiers of the U.S. Third Army in December 1944, shortly before the beginning of the Battle of the Bulge. The prayer for good weather became known, appropriately, as "the Patton Prayer."

I decided to make use of the story of the prayer to bring credit to General Koch for his exceptional intelligence work preceding the Bulge. The timing of events made for an easy connection. On December 8, Patton called on the Third Army chaplain, Colonel James H. O'Neill, to write a prayer for good weather. The following day, Koch briefed his commander and the rest of the staff on the enemy buildup in the Ardennes and warned of the danger of a German counterattack. The attack came a week later.

I combined the two story lines—the Patton Prayer and Koch's superb intelligence work during that period—into a single newspaper feature article that has been published multiple times. It was first published in 1967 and most recently in 2011.

Because it deals with historical fact, the piece has hardly changed over the years. For religious publications, I wrote a version titled "The Miracle of Bastogne."[1]

General Koch had always told me Patton was deeply religious. Chaplain O'Neill substantiated this view. Writing four years later about his meeting with Patton that day, he recalled that the commander expressed a strong belief in the power of prayer. He said Patton asked not only for a weather prayer, but also for a training letter on the importance of prayer to be sent to all the Third Army chaplains and unit commanders down to battalion level.[2]

The chaplain wrote that Patton told him, "We've got to get not only the chaplains but every man in the Third Army to pray. We must ask God to stop these rains."

On December 11 and 12, more than three thousand copies of the chaplain's Training Letter No. 5 were distributed. It advised that, "This Army needs the assurance and the faith that God is with us. With prayer, we cannot fail." In case there was any doubt, O'Neill added his pledge that the letter had "the approval, the encouragement, and the enthusiastic support of the Third United States Army Commander." The prayer for good weather, printed on the back of the card that carried Patton's Christmas greetings to the troops, also was distributed at this time.

The prayer was supposed to be in the hands of every Third Army soldier by December 14. The text of the prayer reads:

> Almighty and merciful Father, we humbly beseech Thee, of Thy great goodness, to restrain these immoderate rains with which we have had to contend. Grant us fair weather for Battle. Graciously hearken to us as soldiers who call upon Thee that armed with Thy power, we may advance from victory to victory, and crush the oppression and wickedness of our enemies, and establish Thy justice among men and nations. Amen.[3]

Chaplain O'Neill sent a signed copy of his remarks, published in the journal, *The Military Chaplain*, to Oscar Koch. He concluded that the prayer fit precisely into Patton's formula for success. In that meeting on December 8, Patton claimed the prayers of "people back home" were a significant reason for Third Army's achievements. Patton said success comes through planning, working, and praying.

He went on to quote Patton's explanation, "Any great military operation takes careful planning, or thinking. Then you must have well-trained troops to carry it out. That's working." But there always is an unknown and that unknown "spells defeat or victory, success or failure." He said, "Some people call that getting the breaks, I call it God.... That's where prayer comes in."

So far as O'Neill was concerned, the Patton Prayer merely enhanced Patton's reputation as one of America's greatest soldiers. "He had all the traits of military leadership," O'Neill wrote, "fortified by genuine trust in God, intense love of country, and high faith in the American soldier."

These were descriptive terms Oscar Koch agreed with.

When I wrote about the Patton Prayer, I went on to pose the question of whether the German advance might not have been prevented. I detailed Koch's briefing to Patton and other staff members at Third Army headquarters on December 9, and concluded:

> The question remains, though, whether the tragic American losses in the Bulge might have been prevented. Koch's intelligence reports had accurately judged the enemy capabilities and predicted the coming course of events. Had Bradley and Eisenhower taken the precautionary steps that Patton took, the Allies could have been waiting and sprung a trap that ended the German thrust before it began.

General Koch appeared to be grateful for the article, especially my efforts to include the story of his pre–Bulge intelligence work. Even without this added feature, though, I think he would have appreciated any effort to portray the Patton Prayer story accurately.

In 1947, *True* magazine had published a fictional version of the prayer which had gained some credence as being authentic. Even the Fourth Armored Association newsletter, *Rolling Together*, in its 1965 Christmas issue, reprinted it under the heading, "The following is General George S. Patton's famous, miracle–working prayer, delivered just before Christmas during the dark days of the Battle of the Bulge."[4]

This fictitious prayer still surfaces from time to time. It begins, "Sir, this is Patton talking. The last 14 days have been straight hell ... I'm beginning to wonder what's going on in your Headquarters. Whose side are You on, anyway?"

Robert S. Allen sent General Koch a copy of the *Rolling Together* newsletter that contained the reprint. On the facing page he scrawled, "The legends grow and grow and grow." The general agreed that any effort to convert such legend to fact was a worthy undertaking.

Given the success of the Patton Prayer story, I began to think about ways to build some advance interest in our book. It was clear that the most provocative part of the completed work would be its intensive treatment of General Koch's sterling pre–Bulge intelligence analyses. I suggested that a magazine article on this topic, specifying that a book was to come, might be useful if I could get it published. The general said he thought it was an excellent idea.

If we were lucky, my article would generate enough controversy to be noticed, a good possibility because the material I had to work with flew in the face of conventional wisdom at the time. I drafted an article I hoped not only would call attention to the book we were working on, but also give due credit to Oscar Koch and explode the myth that no one had expected the German offensive.

The general looked it over, made a couple of small suggestions, and said he hoped I could get it published.

The article emphasized the force of the German offensive and its costs, both immediate and long-term. I noted that both Eisenhower and Bradley had written that they had no reason to suspect the strength of the German reserve force or that it might be used offensively in the Ardennes—the one tissue-thin link in the Allied front.

Had that been true, there would have been ample reason to question the work of the entire Allied intelligence apparatus. But it wasn't true. "A previously unheard voice has now been raised," I wrote. "An unsought source in the past, it may cause historians and military men alike to dig deeper into the story of the Bulge." The voice, of course, was that of Oscar Koch.

General Koch, I said, had been busy. And, "With voluminous files of source materials, an untapped supply of personal experiences and memories, and an indefatigable mind for detail, the general is writing a book. His story will leave little doubt that the intelligence staff of General George S. Patton's U.S. Third Army pieced together and disseminated detailed evidence of the latent enemy preparation in those wintery weeks preceding the Bulge, but that their work went unheeded at higher headquarters."

I went on to detail Koch's intelligence analyses over a period of weeks before the Germans launched their offensive. I used the general's own words to describe his December 9 briefing of Patton: "The purpose of intelligence is to assist the commander in accomplishing his mission and protect his command from surprise. Although the heavy concentration of enemy reserves was well out of Third

Army's planned zone of advance, it would certainly be a threat to our flank. We felt the threat was too great to be ignored."

While I was working on the article, General Koch was hospitalized and underwent more surgery. I had corresponded with David Maxey, managing editor of *Look* magazine, who expressed interest in the manuscript, and I had to write and let him know I'd been delayed in completing an essential final interview. I was able to finish the piece after only a brief delay but, ultimately, Maxey turned it down. He wrote that he saw considerable merit in the article and had held it a long time "precisely because it seemed worth having a number of people read it."[5]

And there were others. John Fink, editor of the *Chicago Tribune Magazine*, said it showed a "masterly grasp" of the subject, but was too technical for his audience.[6]

Probably my greatest frustration came from *Life* magazine, where Edward Kern also found it too technical. But he wrote that, "Your article on the Battle of the Bulge has the air of settling at last a vexed question. If it is what it seems to be, you have indeed made a useful contribution to military history."[7]

Close, as they say, but no cigar.

To my everlasting regret, I was not successful in getting the article published. I queried a number of leading publications and several expressed interest, only to reject it later. Complimentary letters on the importance of the topic notwithstanding, in the end it seemed not to meet any of their needs or interests. I'd experienced this often enough as a free-lance writer, but at this point I was beginning to sympathize greatly with the general and his experiences with his earlier book, *Intelligence in Combat*.

General Koch always kept his friend, Robert S. Allen, informed about what we were doing and sent him a copy of the article. When Allen wrote back saying good things about it, the general smiled broadly as he showed me Allen's letter.

"Bob Allen is not quick to hand out praise," he said.

Sometime later, the general wrote Allen, "The Battle of the Bulge article was returned this week by *Esquire*, with nothing said or done. Bob was rather disappointed to get it back as educational material being in the mails a matter of a week from New York, and rather battered when he had sent return first class postage." He added that this was the first instance in which the manuscript was returned "without the courtesy of a reply or acknowledgment of any kind."[8]

I believe General Koch's objective may have been to impress Robert Allen with my persistence in trying to get the article into print, although we never talked about it. The general took my failure in stride. I suppose he had grown accustomed to not having his voice heard.

Chapter 18

TAKING ADVANTAGE of his improved physical condition, General Koch contributed more time to our work and I put in every hour I could spare on top of my job and class commitments. Although mine was a part–time effort on a limited scale I'd find exasperating today, at the time it somehow felt as if we were making good progress on our grand undertaking. I still was exceptionally ignorant about the amount of work it takes to write a book.

In hindsight, I suspect the general had very little hope that he would live to see *G–2* through to completion. He understood his circumstances much better than I did. He'd lived a full life, had survived grave dangers, and without doubt accepted the fact that his cancer was only in temporary remission and his time was severely limited. He was pushing himself harder than he should have.

I also suspect he was doing everything in his power to save me the apprehension he knew I'd suffer if I understood the truly fragile state of his health. He would maintain a strong front and hold on as long as humanly possible. I am forever grateful.

For my part, I had changed positions at Southern Illinois University, moving into a job that made it a bit easier to find time to work with the general. The move was a short one, a single floor higher in the elegant old Susan B. Anthony Hall administration building, leaving the PR office in favor of becoming editor of alumni publications. This meant producing a monthly magazine and newsletter instead of daily press releases and more freedom in organizing my time.

From my standpoint, our prospects looked rather promising. The general was much stronger physically and I had no doubt we would finish the book and see it in print. Later, I would become much less certain.

There were two chapters in General Koch's original *Intelligence in Combat* manuscript that he wanted to retain, both of them tributes to men he admired and respected and perhaps not fully objective—my assumption, not his. He asked me to edit them. I may have revised a sentence or two, but I took great care to keep the general's words because they gave succinct expression to his views of Patton and Robert S. Allen.

The chapter on Allen was titled "What Makes a G–2?" After reflecting on the personal characteristics of a good intelligence officer, Koch stressed the importance of a G–2 team.

"No one individual could handle all intelligence affairs and provide all the answers to all the questions about all the things which were a direct G–2 responsibility," he wrote. He went on to portray Allen, his assistant G–2 on Patton's Third Army intelligence staff, as an outstanding example of the right kind of man for the job.[1]

Allen, a prominent journalist in civilian life, also had a

strong military record. He was an officer in the Wisconsin National Guard and a graduate of the Cavalry School at Fort Riley, Kansas, and was called to active duty in early 1942 as a major.

The Army sent him to the prestigious Command and General Staff School at Fort Leavenworth, Kansas. Completing that, he ended up in the G–2 Section of the U.S. Third Army while it still was in the States under the command of General Walter Krueger.

Allen was among the stateside Third Army staff officers who, on arrival in England, were melded into Patton's handpicked transfers from Seventh Army in Sicily. He and Oscar Koch enjoyed a good relationship from the outset.

In the spring of 1945, Patton personally chose Robert Allen—by this time a colonel—to lead a reconnaissance mission to check out the story told by a captured German officer. The prisoner, a middle-aged major, had offered information about a series of new enemy communications centers. He said he was a "good German" and because he knew the war was lost he wanted to see it end as soon as possible to spare his country further destruction.

This had become a somewhat common occurrence as the Allied forces moved closer to Berlin.

Colonel Allen verified the prisoner's story. In the process, though, he was wounded and taken prisoner. He was hospitalized immediately and an Austrian surgeon amputated his badly damaged right arm. Five days after his capture, Third Army troops took over the area and set him free. Allen was evacuated back to an American Army hospital and after a total absence of only 17 days was back on full military duty at his desk alongside Oscar Koch.

General Koch never portrayed Allen as the perfect example of an intelligence officer. He pointed out that the Third Army G-2 Section was made up of 16 officers and warrant officers and 25 enlisted men. They included, along with a few Regular Army officers, a high school principal, a grade school teacher, an international banker, a musical instrument salesman, a manufacturer's foreign representative, three journalists, a real estate dealer, a retail shoe salesman, and four lawyers. Two of the lawyers had only recently finished college.[2]

"Pretty diverse backgrounds," I said, looking over his list.

"It wasn't the civilian occupation that made them good intelligence men," the general said. "It was all the qualities I put in the book."

He had written, in the chapter now to be included in *G-2: Intelligence for Patton*, that all of them possessed imagination, initiative, and mental flexibility. They all were willing workers and methodical detail men and organizers, able to work quietly and in harmony with others. None was a worrier, unable to relax. They all could supervise others, and they all could think on their feet and express themselves cogently.

"And they all took orders from you," I said. It was a statement of fact, not a question.

"They didn't need many orders. They all knew their jobs, and they did them. Very well, too."

"But you were the man at the top. You were the one who reported directly to Patton. If one of them made an error, you took the responsibility, right?"

"Yes." And he smiled that modest smile. "But I also got a lot of the credit for their good work."

I wanted to know more about how Patton responded to mistakes. I would have guessed that the offender got a chewing out that took not only his hide but some of his flesh. The general said that was a common misconception.

"The Old Man was human," he said. "He let you know when he wasn't happy about something. But he didn't hold grudges. Not against his own staff. Not against anybody he knew usually did his job. And especially not against anybody who admitted they'd made a mistake."

General Koch's explanation squared with other reports.

Warrant Officer Fred Hose tells the story of Captain Helmut Gerber, a German refugee valuable to Koch's G–2 Section because of his abilities as a translator. One day Gerber was on duty as operations officer and took a phone call. The speaker on the other end of the line said, "This is General Patton and I want to be briefed!" The captain thought it was one of his fellow staffers playing a joke on him and made some flip remark.

Patton was calling from just down the hall. Captain Gerber now heard him without the phone to his ear. "Godammit! This is General George Patton and I want to be briefed." He quickly answered the phone with proper military courtesy, but feared his blunder would lead to his immediate transfer. When Koch returned from lunch, Gerber told him what had happened and expressed his fear of likely results. Koch told him to go to General Hap Gay, Patton's chief of staff, and apologize.

Gerber took Koch's advice and never heard anything more about the incident.[3]

Robert Allen tells the story of an officer who was berated by Patton for something he hadn't done.

Patton later apologized. A friend asked the officer why he hadn't explained the situation at the time. The officer said there were two reasons: "First, 'Georgie' isn't the kind of guy you argue with. Second, I knew that when he found out he was wrong, he would right it."[4]

General Koch sent a draft of the chapter on "What Makes a G–2?" to Allen for fact checking. Allen wrote back that he had found no errors and had made a copy for his files "because of your very moving and profoundly appreciated account of my experience."[5]

As to the chapter on Patton, it was virtually a worshipful tribute. I understood by this time how much General Koch idolized his old commander and certainly I would not have expected anything else. If I had found verifiable factual errors in Koch's manuscript I would have been obligated to correct them; insofar as personal opinion was concerned, he had every right to express it. I know that many would not fully agree with his assessment, but I know, also, there probably was no one who knew Patton better than Oscar Koch did. His subjective evaluation was based on years of close association with "the Old Man."

The general noted that Patton gained a lasting reputation as a pioneer in the use of tanks in combat in World War I, but many junior cavalry officers became acquainted with his name initially through their familiarity with the Patton Saber. He used the saber as a metaphor for Patton himself, noting it was designed to "strike the enemy head-on, driven home at full speed by the horseman—a weapon of thrust."[6]

I asked the general for his explanation of Patton's success.

"Patton stayed on the attack," he said. "He still used

some of the basic tactics of the old horse cavalry. He wanted the enemy on the run and would keep the pressure on until the enemy force was pinned down, and he would try to maneuver around the flank and get behind the enemy and cut them off from the rear."

"Speed was everything," I commented.

Yes, the general said, and he went on to tell me a story about a reluctant corps commander who didn't think he could move his troops as fast as Patton wanted him to. The corps commander had detailed a number of things that had to be done and complained to Patton, "You must remember that it takes time to do these things."

Patton responded, "Then what the hell are you wasting time here for?"

Koch laughed heartily as he finished the story. "The Old Man didn't have much patience for anybody who made excuses," he said.

It did not come as a surprise to me that, in his chapter on Patton, the general tended to come back to the cavalry from time to time. He wrote that Patton was a firm believer in the cavalry motto, *mobilitate Vigemus*—In mobility lies our strength. He said Patton's success often came from calculating the minimum time it would take the enemy to act, and then acting sooner. By progressively following up his first action by a second one, again in less time than that minimum, he would catch the enemy maneuvering to react to the first, and continue this cycle as long as he was in control of the situation.[7]

In an understatement striking in its simplicity, Koch wrote that, "George Patton loved military life." But he drew a clear line of distinction between military life and the ravages of war. When his troops were in combat, Patton believed he should be seen at the front, getting shot at. "He felt it was a great morale builder for the troops to see a commander sharing the danger of his soldiers."

Koch made no bones about Patton's penchant for stern discipline. He insisted that in the long run this had a lot to do with the "pride and spirit evident among the men of Patton's commands." But it was not discipline alone which created "the spirit which permeated Patton's troops, from top staff officers downward. He was a commander who earned the respect of his men, a firm adherent to the philosophy that it was a commander's duty to lead his men, not push them."

Oscar Koch's high regard for Patton and Patton's high regard for him have been acknowledged by military historians, if not as a model of the way this relationship should be then certainly an important reason why they were an extraordinarily effective team. Koch's temperament was such that he probably could have worked effectively with any commander, but the same cannot be said of Patton.

Michael E. Bigelow, in an insightful article on the history of intelligence in the U.S. Army, cites the rapport between Koch and Patton and says that while "the flamboyant and gregarious Patton and the modest and soft-spoken Koch offered a dramatic contrast, they made an effective team." He says that although Koch looked more like a kindly old college professor than one of Patton's closest advisors, "his appearance belied a perceptive mind and a keen understanding of intelligence. These qualities, combined with deep loyalty to Patton, made Koch almost indispensable."[8]

Carlo D'Este—whose *Patton: A Genius for War* I still

consider the best Patton biography—contended that Oscar Koch "was the perfect G–2 for Patton, calm, deliberate and with just the right personality to interact with his volatile boss. I'm sure they each admired and appreciated each other."[9]

Everything I learned from the general leads me to believe these assessments are accurate.

Patton and Koch came as close to a perfect match as can be found in the entire U.S. Army command structure in World War II. I suppose it can be said that, unlike intelligence officers, perfectly matched commanders and G–2s are born, not made. This surely was true in the case of George S. Patton Jr. and Oscar Koch.

Chapter 19

THE STORY OF Oscar Koch's service in World War II was dramatic to the end. Poring through the records of his work as G–2 for the U.S. Seventh Army in Patton's Sicilian campaign and Third Army G–2 across France and into Germany had been a fascinating activity. Relative to the chronology of his wartime activities, we were nearing completion.

My feelings were mixed. On the one hand, I was eager to finish our manuscript. On the other, it was as if I were reading a thrilling adventure story that I hated to see come to an end.

My work with General Koch, now well into its second year, had served to blunt to some extent my concern for what was happening in the world around us. The war in Vietnam ground on, while the government got tough with draft dodgers. The year began with the indictments of Dr. Benjamin Spock, the noted pediatrician and writer, and the Reverend William Sloan Coffin Jr., the Yale chaplain, for aiding and abetting draft defiance.

Coffin came to Carbondale not long afterward to deliver a speech at Southern Illinois University.

There was immense interest and curiosity on campus about what he might say. I went to a packed Shryock Auditorium to hear him, and I remember telling a friend afterward that he gave a perfect "swords to plowshares" sermon.

The Tet Offensive by Vietcong and North Vietnamese forces and the costly Battle of Khe Sanh gave impetus to the antiwar movement. Meanwhile, the defense department called 24,500 reservists to two years of active duty to help meet the need for more troops. Minnesota Senator Eugene McCarthy, running as an antiwar candidate, mounted a vigorous campaign to oppose President Lyndon Johnson's nomination for another term in office. With Robert Kennedy also in the race, Johnson would soon announce his decision not to seek reelection.

It would be some months yet before the antiwar movement built to its pinnacle, but the deaths of three college students in a civil rights protest in Orangeburg, South Carolina, had stoked the fires of racial conflict and the April assassination of Dr. Martin Luther King Jr. in Memphis brought racial tensions to a boiling point. President Johnson took to radio and television to urge Americans to avoid violence, but to no avail. The Associated Press reported that throughout the country during the night "crowds of angry Negroes broke windows, looted stores, threw firebombs that started many blazes and attacked police with guns, stones and bottles."[1] Several deaths were attributed to the violence. The riots continued for several days.

The night of King's assassination, someone painted "The only good nigger is a dead nigger" on a thoroughfare sidewalk on the Southern Illinois University campus. This was an obscene incongruity at an institution that had a significantly higher minority student enrollment than the national average. Nonetheless, given its location, the crude epithet was seen by hundreds or perhaps even thousands of pedestrians before groundskeepers could assemble the wherewithal to clean it off the concrete.

Later in the same month, Robert Kennedy was assassinated in Los Angeles. And it hadn't happened yet, but in the fall clashes between antiwar protesters and police in Chicago would mar the image of the Democratic National Convention where Hubert Humphrey was nominated to replace Johnson as the party's candidate for president.

Although General Koch and I rarely talked politics, there was one political figure that he disliked too much to ignore. This was Curtis E. LeMay, the retired Air Force general who was a candidate for vice president in 1968 on the Independent Party ticket with former Alabama governor George Wallace. LeMay had been head of the Strategic Air Command and served as U.S. Air Force chief of staff, but he perhaps was best known for a comment about America's ability to bomb the North Vietnamese "back to the stone age" as a way to end the long–running war in Vietnam. This statement, from which LeMay never backed away, was widely reported and came to identify him in the minds of many.

I arrived at the Koch home late in the afternoon just as the general finished reading a newspaper report on the press conference where LeMay was introduced as Wallace's running mate. The report said LeMay attacked what he called America's "no will to win" policy in Vietnam and stated his philosophy that, "When you get in it, get in it with both feet, and get it over with as soon as you can." And then he added, "I think there are many cases where it would be most efficient to use nuclear weapons."[2]

General Koch shook his head in disbelief.

"His saber rattling should scare people to death," he declared. "Where do we get nuts like this?"

I would have liked to talk more about the Wallace–LeMay ticket. The general had other ideas. He had a big stack of new material for us to work on, and if I was ready perhaps we should get to it. For a general, I replied, he gave very mannerly "orders."

"So we finished the Battle of the Bulge," I said. "Where do we go next?"

We plunged into his pile of notes and documents, with him explaining what he had and what it meant and me taking notes frantically as I tried to keep up. The story—his story, as much as I could make it so—had lost none of its appeal.

Once the Battle of the Bulge was over, Patton's Third Army moved swiftly into the German heartland. German soldiers surrendered by the thousands. By mid–March 1945, with well over 185,000 prisoners taken during the past eight months, the actual prisoner count varied from Koch's early intelligence estimates by only 518. I found this remarkable, but he insisted that it was routine "once you know the enemy's orders of battle." Good intelligence work, he said, would take it from there. I still found it remarkable.

POWs always were a potential source of vital information, in the Third Army G–2 Section as elsewhere. One study showed that more than a third of all World War II combat intelligence information came from POWs. This was attributed in part to the great number of Germans taken prisoner and also to the fact that four Army interrogation teams could interview as many as five thousand prisoners a day.[4]

General Koch explained that prisoner interrogation normally began at a lower level, tactical questions being asked by battalion, regiment, division, and corps specialists in POW interrogation. Only if something big turned up would a captured enemy soldier likely end up at Third Army headquarters.

One German prisoner who did make it to that level for questioning had information that was important in part because of what he did not say. He had a map on which were marked the precise locations of important military installations. These were checked out over time and proved to be an accurate accounting. But of even more immediate interest, the general pointed out, was the fact that the prisoner's map "showed nothing in the area where Hitler's last-stand mountain fortress was supposed to be."

High-level Allied thinking in the spring of 1945 gave a high priority to the possibility that Hitler and his inner circle would make a determined last stand in a nearly impregnable fortress in the Bavarian Alps, commonly referred to as the Alpine Redoubt. German propaganda had vigorously promoted this, perhaps as a rallying point for the faithful Hitler followers.

"Either to accept or reject this last minute threat was important to future thinking and planning," General Koch said. "The Third Army G–2 Section had set out to determine whether or not the Redoubt actually existed. We started with a theory that because of the suspected location, deep in the mountains in a region where there were few roads, the hideout would have to be supplied well in advance."

This meant that a careful analysis of information on German troop movement in that area would be a good place to begin. Koch's team undertook a detailed study of the movements of all SS divisions—Heinrich Himmler's *Schutzstaffel*, Hitler's most fanatic followers—assuming these units were most likely to be among those guarding the mountain fortress if it indeed existed. They found that between January and mid–April, not a single SS division had been moved from another area into the region in question.

As we recorded their findings in *G–2: Intelligence for Patton*:

> One SS mountain division (Andreas Hofer) had been in the Austrian–Italian border area between Innsbruck and Bolzano astride the Brenner Pass for months. The 24th SS Mountain Division was reported forming northwest of Trieste, and the 14th SS had been reported moving from Czechoslovakia in March to an area near Ljubljana, Yugoslavia. The 16th SS Panzer Division had moved from below Venice to the Czech–Polish border in February and to the south of Graz, Austria, in March and April. No others showed any indication of being available for the occupation of the mountain fastness of the Redoubt. Tactical air reconnaissance found no unusual motor or rail activity supporting reinforcement of the area.[5]

Third Army G–2 reports for the period mentioned the Redoubt in discussions of the general enemy situation, but did not list it among enemy capabilities. "We labeled the Redoubt a myth," Koch said. "It had been built up in talk, but not in fact."

Patton, once again showing confidence in his intelligence staff, personally discounted the existence of the Redoubt, even though he was ordered to move forces to the south to cut off any German movement in the area. And once again, time would prove his confidence well justified. If Hitler ever had any such plans, they never came to light.

The Third Army intelligence staff, using the same intelligence available to all higher Allied commands, was the only one to correctly write off Hitler's Alpine Redoubt as a mere fabrication.

The general never boasted about such accomplishments. Differing intelligence conclusions at different headquarters were not unusual, he explained. Intelligence sections issued written summaries of enemy capabilities, with one capability that was "favored" in that headquarters. The reports included discussion that gave the reader the background reasoning and the logic that went into its choice. Each headquarters thus reached its own decision, even though the intelligence information on which the decision was based might be the same as that considered by other units.

"In the Third Army G–2 Section," Oscar Koch said, "if we had time we wanted every one of the senior intelligence staff members to lay out his thoughts and ideas for open discussion. We wanted all opinions to be heard. It was always a team activity."

"Patton's intelligence team served him very well," I said.

"We did our best. We wanted to give him reliable information, and I think most of the time we did."

Patton apparently agreed with this assessment. On May 7, 1945, the surrender of all German forces was announced, effective at one minute after midnight May 9. On May 8, Patton received word that the commanding general of German Army Group South wanted to discuss terms of surrender. Patton delegated the responsibility to Koch.

"The Old Man said, 'You go meet him. You know him better than I do. You've been dealing with him a long time.'"

A high honor, I observed, but an appropriate one. The general offered no response, but his smile said a great deal.

With the end of combat in Europe, Patton's forces became an army of occupation. Patton soon was reassigned to command the Fifteenth Army as punishment for a press conference analogy comparing Nazis in Germany to members of the party in power in the two-party political system of the United States. Not long afterwards, he was gravely injured in an automobile accident and died from his injuries.

I hated to bring up the topic. There was no question General Koch would find it difficult, given his respect for and devotion to his old commander.

"Were you there at the time of his accident?" I asked softly.

"I was back in the States," Koch said. "It was my first break since we started planning for the North African invasion. I would have rejoined him soon. I was with him a long time." Tears came to the general's eyes and his voice wavered.

"You're a Patton man," I said.

"Yes. And I always will be."

After Patton's death, Oscar Koch was sent back to Fort Riley, Kansas, to head the intelligence department at the U.S. Army Ground General School, the first peacetime intelligence school in Army history. Here, at last, was his opportunity to make use of his long-held conviction that intelligence officers are made, not born. Robert S. Allen would contend that this appointment was an even greater tribute to the Army's intelligence than it was to Koch.[6]

One military historian has observed that this was a heady time for the intelligence officers, former wartime S-2s (lower echelon intelligence officers) and G-2s, who assembled at Fort Riley. They felt they had a lot of lessons to pass along—lessons they'd learned the hard way.[7]

Back at Fort Riley, Koch soon was detailed to gather furnishings for one of the post's old buildings which was about to be rededicated as Patton Hall, a memorial to his old commander. We never discussed it, but one can well imagine that he took immense pride in fulfilling this mission and found particular satisfaction that "the Old Man" was so honored on the site of the Army Cavalry School where they had first served together.

The cavalry itself, though, had seen its final days. Some twenty years after the fact, General Koch would recount his role in what he referred to as "the passing of the cavalry."

He wrote that General Jacob L. Devers, then commander of U.S. Army ground forces, was at Fort Riley to deliver a commencement address. While there, he directed Koch to write a letter that would justify government funding for the horse cavalry. The letter was to defend a detailed budget for "the horse, horsemastership and horsemanship arts and techniques" but should not include dollar amounts. These would be developed later.

The letter must be succinct—no longer than a page and a half, single-spaced. The letter was to be prepared for General Devers's signature.

General Koch gave a clear impression of the careful work that went into the letter. He and the top-level members of his academic staff worked on it for several days, until every word had been "agreed upon as understandable and essential." It justified "so many remounts per year, so many equitation instructors, so many saddlers, horseshoers, farriers, etc., etc., as well as students, all at a bare minimum basis to be economical as well.

"All the whys were given, and all soundly supported. A bare nucleus for the United States Army in the event that one day again, these fundamentals would become factors in our national defense schedule. And all of this about 1947 or 1948."[8]

The letter itself did not go into such necessary accessories as stabling, straw, oats, hay, building maintenance and the like, even though the team that drafted it assumed these would have to be accounted for somewhere along the way. The same assumption was made regarding rations for officers and men involved as well as housing costs.

Oscar Koch heard nothing on the fate of his efforts at the time. It was only after he was sent back to Europe in 1949 as chief of intelligence for the commander of American forces in Austria, General Geoffrey Keyes, that word surfaced that the horse cavalry had just about reached the end of its days and "the horses and other impedimenta had to be disposed of." The rest of the story in the general's own words:

Somewhere along the line Peter C. Haines III, then a major general, I believe, and I were visiting. He had, as I recall, been one of the liaison officers between the Army and the Defense Department, with the Bureau of the Budget, or closely related thereto—so closely that it would be he who would defend the budget before the Secretary of War in the preliminary arrangements as to what the Army budget would be as presented to the Congress for approval.

They sat around a table, Haines ready to speak up if and when asked to. But relating to this important item there were no questions, and hence no answers. Mr. Secretary had a red pencil in hand, and in going through the pages item by item, with one bold stroke lined out "Forage," and with it went the Cavalry: no food, no horses; no horses, no stables; no stables, no riders; no riders, no personnel pay, no nothing. No Cavalry. No arguments. That was it.[9]

The way Oscar Koch loved the old horse cavalry, he must have been extremely disappointed at its inglorious end. I remembered my astonishment when I learned that Patton's Seventh Army troops used horses in Sicily, and the way the general brought it up. We were talking about some aspect of intelligence operations when he paused, as if interrupting himself, and announced, "By the way, it wasn't the cavalry, but we used horses and mules in Sicily."

I expressed my surprise. "Seriously? A mule is a pretty slow means of transportation."

"It was mostly for pack animals. The terrain was so rough in some places that pack animals were the only way you could carry supplies. We captured some mules from the Italian army, and also bought some from the local Sicilian residents."

The general went on with other stories. I wanted to know more about the pack animals, but I didn't want to bother him with questions about the mules. I made a note to myself to do some research on the topic later. When I got back to it, I learned that Seventh Army used about four thousand pack animals, mostly mules. Units procured civilian animals by having local police in towns along their routes of combat call for the owners of horses and mules to bring them in for examination and potential lease to the Americans.

The Army rented those animals selected at a standard rate of fifty lire a day. In addition, mules were valued at $150 and horses at $120, to be paid if the animal wasn't returned. Mares, however, were valued at "$10 less than the price of male animals." The owners painted numbers on their animals or branded their hooves for identification. The animals were important, carrying 250– to 275–pound packs of water, rations, radios, and ammunition over the mountainous Sicilian terrain. It was dangerous duty. Some fifteen hundred were lost in action. The rest were returned to their owners at the end of the fighting.[10]

Chapter 20

OSCAR KOCH BEGAN his assignment in Austria as chief of intelligence for General Keyes at a convenient time relative to one of the more controversial episodes in post–war European intelligence operations. At the end of the war, General Reinhard Gehlen, a senior German Army intelligence officer and an expert on the Soviet Union, directed his men to preserve their records and surrender to the American forces. Gehlen and several of his officers were sent to the U.S. in the summer of 1945 for debriefing, after which—with the Cold War germinating—the Army began to use them as operatives in the American occupation zone in Germany.[1]

From 1945 to 1949, the U.S. Army handled contacts with what came to be known as the Gehlen Organization and funded its intelligence collection activities. The CIA, after a long debate on the matter, took over that operation on July 1, 1949, the year Koch arrived. This likely means he had extensive contact with the CIA during his years in Austria, but time ran out before we were able to talk about it.

It was apparent the general's physical condition had

taken a sudden downturn. We had reached that point where I finally accepted the fact that if he was to have time to fully review the finished manuscript for *G–2: Intelligence for Patton*, much less see it through to publication, it had to be wrapped up quickly.

General Koch actually took some satisfaction in this; he wanted the book to be about combat intelligence, not the cloak and dagger stuff the CIA was noted for. But he let me know that however I chose to bring our work to a conclusion would be acceptable to him.

I wanted to include at least skeleton information about his assignment with the State Department in Washington, D.C., which followed his service with General Keyes. I'd learned from John W. Allen that this period involved work for the CIA, and I was eager to find out more. I didn't get very far. Koch was brought into the position by General "Beetle" Smith, the former Eisenhower chief of staff who became CIA director in 1950. Smith is recognized for restructuring the agency and making it more effective. And clearly it is to his credit that he chose in Oscar Koch one of the most brilliant minds available when it came to anything that had to do with intelligence.

When I asked General Koch about his role there, he shrugged off the question with an assertion that he "didn't do anything very important."

"Did you work directly with Beetle Smith?"

"Not really. One of us would work in the daytime and the other at night. We just met in the hallway in passing to and from work. It was actually kind of silly, the way we acted—just nodding as we passed, like we barely knew one another."

"Because you didn't have a good relationship?"

"No, no. When I say it was kind of silly, it was because we behaved like we were involved in some kind of covert activity. Like it was a secret that we both worked there or something."

I tried to find out more about what Koch actually did. All I got was generalities about collecting intelligence information. I take him at his word.

General Smith certainly knew there was no one better at the nuts and bolts of collecting and analyzing intelligence than Oscar Koch. If Koch's contributions to the CIA were based on this, I've no doubt it was a worthy role he played.

It was during this stretch in the State Department that the general ran afoul of the interests of his cousin, Dr. Murray Zimmerman. My information on this comes from Dr. Zimmerman, not Oscar Koch. Dr. Zimmerman brought it up as the single instance in which, in his view, Koch behaved in a mean-spirited way. I didn't tell Dr. Zimmerman, but after getting the full story I found it quite easy to side with his cousin Oscar. I found it a story worth passing along.

After the war Dr. Zimmerman, who had lived and studied in Germany, became friends with Colonel Otto Skorzeny, Hitler's commando leader. Skorzeny had directed raids on Allied forces and was best known for his daring rescue of the imprisoned former Italian dictator, Benito Mussolini.

Skorzeny was burdened with the ugly sobriquet "Scarface" because of a long dueling scar down the side of his face and across his mouth. His wife, Ilse, had asked Dr. Zimmerman if he could perform plastic surgery on her husband's face and the doctor agreed, provided he could get Skorzeny to the U.S.[2]

Skorzeny had been tried as a war criminal and, even though he was acquitted, still was not welcomed in the U.S. Dr. Zimmerman appealed to his cousin Oscar Koch in the State Department, hoping he would help get Skorzeny a visa. Instead of cooperating, though, Koch exploded, "I wouldn't do anything to help that son–of–a–bitch."

Dr. Zimmerman held no grudge against Skorzeny, rationalizing that he simply had been a soldier fighting for his country and saw no reason why Koch should still be angry. As late as 2001, some six years after Skorzeny's death, the doctor was planning to visit Ilse during a visit to Europe but learned she suffered from Alzheimer's disease and wouldn't recognize him.

Skorzeny's biographer, Glenn B. Infield, describes the German commando leader as a "very easy man to talk with, a very interesting conversationalist." But Infield didn't buy into the argument that Dr. Zimmerman made, that Skorzeny was simply fighting for his country—in other words, merely following orders as a good soldier should. Rather, he offers a persuasive argument that "Hitler's commando" firmly supported most of the Führer's political views and engaged in international terrorist activities after the war.[3]

I have no doubt it was General Oscar Koch's reverence for fair play that left him unforgiving of Otto Skorzeny, regardless of post–war activity. The general had a great deal of respect for the regular German Army soldiers, but not for the Nazis or members of the *Schutzstaffel*, the notorious SS.

He believed that warfare, no matter how brutal, should be played by the rules. One of the few comments he made about my draft of a magazine article on the Battle of the Bulge was, "Don't forget Malmédy. People must remember the atrocities." Malmédy, of course, is where the Germans massacred more than one hundred American soldiers from Battery B of the 285th Field Artillery Observation Battalion after they had surrendered.

Colonel Skorzeny's English–speaking commandos infiltrated areas behind the American lines in the days preceding the Battle of Bulge, dressed in American and British uniforms and driving American vehicles and Panther tanks disguised to look like Shermans, with orders to do whatever they could to disrupt the Allies' ability to respond to the coming German offensive. This was to entail the seizure of one or more bridges over the Meuse River and the creation of confusion among the American troops by giving false orders and fouling communications. Even small acts such as removing road signs, cutting telephone lines, and blowing up Allied ammunition dumps could have a major effect.

Skorzeny said his orders came directly from Hitler during a meeting on October 21, almost two months before the offensive was launched.[4]

Oscar Koch would have pointed out that soldiers who engaged in warfare wearing enemy uniforms were considered spies under international law, and subject to execution. In other words, they weren't playing by the rules. The OSS men who worked with Koch's G–2 Sections in Sicily and, later, on the continent, dressed in civilian clothes.

Colonel Skorzeny acknowledged the international law, and had accepted the advice of a higher ranking officer that his men wear German uniforms under their American or British ones and take off the latter before they used their weapons. He also claimed that Hitler, in assigning him the mission, claimed the enemy "has already done us a great deal of damage by the use of our uniforms in various commando operations."[5]

Skorzeny's trial concluded in dramatic fashion when Wing Commander Forrest Yeo–Thomas, the British commando leader known as "The White Rabbit," testified that he and his men had, in fact, engaged in armed combat "many times" dressed in stolen German uniforms. He told the court the British Secret Service "often wore German uniforms, were always armed, and, when trapped, used their guns without hesitation."[6]

Apparently, though, none of this mattered to General Koch. Skorzeny no doubt had been responsible for the deaths of American soldiers. Oscar Koch was not going to show Hitler's top commando any favors, no matter the views of his well–liked cousin, Dr. Zimmerman.

The general never mentioned Skorzeny to me, which I don't find surprising. He rarely brought up the names of those he didn't like. I believe it made him uncomfortable to speak ill of anyone. Surely it was difficult for him not to be critical of Eisenhower and Bradley, and particularly their costly decision to shut down Patton's offensive in favor of Montgomery's Operation Market Garden and their complete and costly failure to heed his dramatic intelligence estimates in the weeks preceding the Battle of the Bulge.

The exception cited earlier, General Curtis LeMay, didn't come up again in our conversation. I was a political junkie who would talk politics at the drop of a hat, but beyond his contempt for LeMay I never sensed any interest in the subject on the part of the general. He apparently was thoroughly grounded in the military tradition of staying non–political.

But even though we didn't talk politics, we talked

about a great many other subjects not directly related to our work at hand. These often grew out of military topics, of course, and no matter how hard we tried to avoid this we found it inevitable that at times we got into discussions that took us far afield.

Though we had talked about it before, the subject of the general's transfer from Sicily to England to rejoin Patton in his new Third Army command came up again in a new context. He had made this somewhat unsettling trip aboard a C–47 transport plane.

"I rode in a C–47 once," I told him. "Actually, I rode in both a military C–47 paratroop carrier and the civilian version DC–3 passenger plane."

I mentioned this partly to make conversation and partly because it was a rare occasion on which I had a story that might be even remotely comparable to his experiences. But his face came alive with the innate curiosity I'd become familiar with. This surely was rare for someone as young as I was, he said. How did it come about? I said there was nothing of importance to my story, but he sincerely wanted to know.

I told him the C–47 was a South Carolina Air National Guard plane. When I was stationed at Fort Jackson I rode it to a football game.

"Say that again. A football game?"

I explained that the Fort Jackson football team had a game at Camp Lejeune, North Carolina, against a team of Marines from that base. The Fort Jackson post commander decided he wanted fans there to cheer on our team and sent out a memorandum inviting each regiment to choose two people for a flight to Camp Lejeune for the game.

"Did this happen often?" the general asked, obviously

eager for an explanation.

I said it was the only instance I was aware of and added, "I think it had something to do with the fact the game pitted the Army against the Marines."

"And you were chosen?"

I was not really chosen, I explained. It merely was a case of nobody else in the Third Training Regiment wanting to go. Rank got first choice, of course, and no one in regimental headquarters showed any interest until it got down to Sergeant Mason Sykes and me. We just said 'thank you very much' and crossed our fingers they wouldn't send it on down to the battalions, and they didn't.

I went on to tell him that on game day the two dozen or so Fort Jackson personnel who were to make the trip drove the few miles to the Congaree Air National Guard base, where a C–47 was waiting. Most of the others were officers, including a major and a few colonels. When the airmen said they had to fit us with parachutes, one lieutenant colonel almost fainted.

"He was that scared?" General Koch asked.

"He looked scared to me. But he wasn't going to admit it. He pulled himself up tall and straight and let them strap on his 'chute."

The general by this time was well into my story. I think this was one of those occasions when he relished talk of things not war or combat related.

"You weren't afraid—with the parachute and all?"

"Not then," I replied. "I was excited to get a chance to fly. But after we were in the air and I saw we were flying over the Atlantic Ocean I had some second thoughts."

I described the C–47's seating, which was built–in aluminum benches parallel to the sides of the cabin, facing inward. We sat on our parachutes, backs to the wall, like paratroopers. The flight to Camp Lejeune and the return trip went without incident.

"Who won the game?" the general asked.

I didn't remember.

Then he asked about the passenger DC–3 I had mentioned. Surely I wasn't old enough to have seen one of those in service as an airliner?

No, I told him. The one I flew on was a corporate plane General Motors had donated to Southern Illinois University. It had an elaborately and comfortably outfitted interior, but after flying on jets it seemed terribly slow. I only rode on it one time.

"I have a flying story, too," the general said.

"You mean a different one?" I asked, referring back to his flight to England.

"Yes. My first flight. In France, during the Great War."

"World War I?"

"Yes. They called it the 'Great War.' They had no idea what was coming."

I couldn't wait to hear his story. I said I thought planes were pretty flimsy back then, and he nodded agreement and laughed.

"The one I was on sure was. It was my first plane ride and it could have been my last."

He said the plane side–slipped into the ground from an altitude of three hundred feet. Although neither he nor the pilot was seriously hurt, the crash left them dangling upside–down in the wreckage. They had to be chopped free with an axe.

"That could discourage you from wanting to fly again,"

I said. "Did it?"

"No, no. It's just like falling off a horse. You get right back in the saddle and try again." He still had the old cavalryman spirit.

The general fell silent. He looked tired. We had covered most of the material we needed to cover, and I had a great deal of copy to draft. I thought a period of rest would do him good. I asked to call our session to a close and promised to get back soon with material for his approval.

"There's one more bit I want to work on," he said. "I want to end the book on 'Command Support' unless you have something else in mind."

I said of course, I still was only the city editor. The story was his to tell.

"I've worked on it some. We were lucky. We always had it. under Patton. Command support means—"

"I know what it means," I interrupted him gently.

"I'll have it finished in the next day or two. You can pick it up, or I can drop it off at your house."

I assured him this didn't matter. Just let me know. A couple of days later he called and said he was finished with the section and would bring it by in the evening.

"It's short," he said. "Maybe it's just an epilogue."

I read it as soon as he brought it, and I knew it was a perfect way to conclude our book. It proved the validity of his contention that he was fortunate to have served under Patton. The first sentence read, "What the intelligence officer needs most to help him through his day-to-day chores is command support: the support of his commander, evidenced primarily by mutual confidence engendered by and nurtured through respect."[7]

From what I had learned through working with Oscar

Koch, this description fit his relationship with General George S. Patton Jr. almost perfectly. They had respected and trusted each other, admired each other, and from all outward appearances the two of them might have been viewing their world through the same lens.

I suddenly understood why, for so many years, the opinion of military historians—their inexplicable failure to give him credit where it was due—was of no great importance to Oscar Koch. He had received credit from the source that mattered most, his commander. He had command support.

"With command support," he had written, "G–2 will tackle any job. Without it, he performs a purposeless task, merely going through a series of staff exercises. In that case, both he and the commander are losers."

The general went on to describe how intelligence was always given top priority in Patton commands: "On many occasions the commander's group included but two others, one of them G–2."

After he dropped off the copy for what turned out to be the epilogue for *G–2: Intelligence for Patton*, it would be three weeks before I saw him again. Realizing, I'm sure, that his time was limited, he had scheduled a trip to California to visit his two sisters there and his cousin, Dr. Murray Zimmerman. I never heard much about his trip, but he came home with a beautiful little white poodle for Nan. The dog's name was Charlene and from that day forward I rarely if ever saw Nan without Charlene on her arm or at her side.

From this point on, the general's physical condition continued to decline rapidly.

Cancer is a malevolent disease that chooses its victims

indiscriminately. It may strike suddenly as a devastating force, or it may advance slowly and deliberately, eating away at life in miniscule nibbles until the body surrenders to frailty and loses its will to fight. It has a way of hiding itself until there is no option for treatment, or pretending it has been beaten merely to lie in wait to return at another time and rip away the walls of hope erected in its apparent absence.

In General Oscar Koch's case, cancer struck hard at the outset, granted a period of remission that let him do the things a dying man would want to do, and then felled him with shocking quickness. I have no doubt he was in great pain at the end. We would never know, because he would not complain.

He underwent chemotherapy that left him terribly sick. He was hospitalized and, at one point, out of touch with reality. He believed he was being held against his will and cleverly managed to call police and ask for help. He cursed the medical staff in German—behavior so entirely out of character that he was humiliated when they told him later what he'd done. This had to be one of the most difficult times of his life.

As long as things were in focus, however, he never lost interest in progress on our book. It was the first thing he asked about when he got home from the hospital. Determination marked his ashen face. Despite his weakened condition, he made clear he was ready to see our manuscript through to the end.

I suppose my own determination was compounded by his. I vowed again that, if it was within my power, I would finish his book and see it in print.

Chapter 21

WHEN OSCAR KOCH retired in 1954 after nearly four decades of service in the U.S. Army, he received a framed scroll signed by General John A. Klein, the Army adjutant general. The scroll was presented in formal recognition of the service he gave to his country. It outlines the highlights of his career and lists the medals he received, including the Distinguished Service Medal, the Legion of Merit, the Bronze Star, the French Legion of Honor, the Russian Order of Fatherland, the Belgian Order of Leopold, and the Luxembourg Order of Oaken Wreath.

Somewhere in my military record rests an acknowledgment that I qualified for the Good Conduct Medal. No one ever told me, and I never received it.

The general, still energetic at the age of 57 when he retired, had an impressive resume. It noted his willingness to accept a job in any geographical area, though he would prefer Europe if the position involved an international location. No objection to occasional travel. He offered leadership skills "inherent in professional Army officer's career" and a working knowledge of the German language. His list of positions held was imposing.

At the same time, though, he applied for a Guggenheim Fellowship to support his work on "A book dealing with military Intelligence in combat, its processes, techniques, methods and concepts and objectives."[1]

He noted that he'd begun this project immediately after his retirement a month earlier, and with support of a fellowship he could continue to live in the Washington, D.C., area "where the Army's historical files, mapping and other research facilities are located and have been made available to me."

The general stated that his sole purpose in undertaking this effort was "to make a significant contribution to the national defense, and I feel that my active service of over 33 years as a regular officer in the U.S. Army fully qualifies me to fulfill this mission." He could have added the additional years he spent as an enlisted man, during which he also had a sterling record.

He received the Guggenheim Fellowship. Work on his book proceeded, even though he continued to do work for the State Department through special assignments in Europe. Some of his correspondence with publishers as his manuscript neared completion originated from posts in Germany and then London, while one letter bore the notation, "Mailed from Paris." I have no doubt these assignments were with the CIA, although I have no evidence to support this.

Unlike General Koch, I did not find a home in the Army. The regimented life is not a life I ever could come to relish. My active service ended much as it began. In my final few months at Fort Jackson, by virtue of time in rank I moved from living "quarters" in the middle of the open second floor of my barracks to sharing a first-floor cadre room with my friend, Sergeant Mason Sykes. He had a portable record player and many nights we lay in the dark and listened to the mellow vocals of Nat King Cole. It was as if I had come full circle, trading the cold winter nights at Fort Leonard Wood and the music of Bobby Hackett for the sweltering Fort Jackson temperatures and the songs of Nat King Cole.

My time at Fort Jackson and my time at Fort Leonard Wood were spent under drastically different circumstances, though. At Fort Leonard Wood, I was a trainee. First the rigorous combat infantry training, then administrative school. No braid allowed on my cap!

Memories from those weeks of training revolve around such things as the unfavored dismounted drill and the stress of early morning wakeup for KP detail. The most consistent consequence of dismounted drill was a whack on the helmet from the rifle barrel of the squad–mate in line behind me when right– or left–face was called. Not the brightest star in the universe, he often turned the wrong way so that the M–1 rifle over his shoulder severely transgressed on my space.

I remember the long hours on KP and how the mess hall cooks would slice up lemons at the end of the day, boil them in a big aluminum pot, and slosh the mix onto the concrete kitchen floor as mop water to cut the buildup of grease. And cleaning the grease trap. I wasn't big on KP.

There was one work detail I was happy to volunteer for. This was the cleaning crew for the bachelor officers' quarters. There was little more to the work than sweeping and dusting one or two of the small apartments, and then I could sit around for the rest of the day in warm comfort and read magazines. My company, meanwhile, was out in the bitter winter cold doing fun things like bayonet drill.

This is not to say I missed all bayonet drill. I didn't. I still remember the spirit of the bayonet: "Kill, kill, kill!"

I remember an instructor, a foul-mouthed old career sergeant who knew his material inside out, who had a speech impediment that made it hard for him to pronounce the "th" sound. We hunkered down and struggled not to laugh when he talked about "sticking your tumb" into the breech of an M-1 rifle. This recollection is so lucid I used it as the basis for a minor scene in my second novel, *The Life and Death of Lizzie Morris*.[2]

Once I was settled at Fort Jackson, life was good. I had great freedom. When my work day was finished—a day of nothing more rigorous than sitting at a desk and hacking at the keyboard of an old Underwood typewriter—I was free to leave the base and live like a civilian. I got a weekday afternoon off because I had to work on Saturday morning. Living like a civilian was hard on $78 a month, half of which went to car payments, but it didn't matter. Mary was there and I saw her almost every night. Sometimes we teamed up on her homework, but on the whole I was just a nuisance who severely distracted from her promising academic future.

I did worry about getting orders to Korea, not only because it was a place I didn't want to go but also because I didn't want to leave Mary. No one told me until just before I was due to be released from active duty that I had been ineligible for overseas assignment the whole time. Why? I had a critical MOS, military occupation specialty. It seemed that administrative clerks were in great demand stateside.

While I was at Fort Jackson, I cast my first vote for president. The minimum voting age still was 21 then, of course, and I'd been eager to vote for Dwight Eisenhower ever since I had campaigned for him as part of a high school civics class project. It's my opinion today that Ike was a better president than military leader.

I benefited from my years in uniform in many ways, some large and some small. Most important, it was because of the Army that I met Mary Corley. And secondly, had I not had the experience of being under arms myself the privilege of collaborating with General Koch on a book that became almost as important to me as it was to him would have been a much more intimidating undertaking.

The timing of my service could have been better. I missed the GI Bill by six months. A retroactive law which was passed later eventually let me get all the money I was eligible for by going to graduate school, but my three undergraduate years at Southern Illinois University was a very uncomfortable period financially. Fortunately, an Illinois Veterans Scholarship covered my tuition.

When I registered for classes the first time, I paid my fees with a $100 bill that was part of the pay I had coming for unused leave when I left Fort Jackson. It was the largest bill I'd ever seen.

Mary worked in a downtown dress shop and later behind the candy counter at an F. W. Woolworth store at a mall. I got a job clearing tables and washing dishes at a university cafeteria. The job paid sixty cents an hour but had one important fringe benefit: Sometimes they had food left over and we were allowed to take this home. Their kitchen turned out delicious pecan pie.

We tried to get by without a car, but Carbondale had no mass transit system and even though it was a small town it was difficult for Mary to get to work. We bought a 1947 Chevrolet coupe for $75. It was hard to start on cold mornings and at times I set an alarm clock to go off every two hours during the night, got up at each alert and warmed up the old Chevy, then went back to bed.

Later, after I got a salaried editorial position on the student newspaper, *The Egyptian*, we were able to afford a better car and paid $100 for a 1949 Plymouth. A day came when the only money I had for gasoline was the silver dollar Walter Young Sr. had given me at the draft board for good luck. It bought some three gallons. This recollection is painful to me today, but at the time I felt I had little choice. The silver dollar I'd kept so proudly was the only money we had.

A draftee enlisted man in 1955 faced two years of active duty, two years of active reserve, and a final stint of inactive reserve. Active reserve required monthly training sessions with a local Army reserve company, which proved a challenge while I was in college. Area reserve units usually were at full strength and didn't much care to take in a bunch of college guys whose official assignments were to home units elsewhere. We went to any company that would take us, for as long as it would let us come. If we missed too many meetings there was a serious threat we'd get orders to summer training camp.

My assignment to the Third Training Regiment at Fort Jackson paid off here in an unexpected way. Fort Jackson in general and the Third Training Regiment in particular were infantry units. My brass insignia indicated that I was in the infantry, even though as an administrative clerk my permanent Army branch was the Adjutant General's Corps. These reserve units always had a stack of administrative work and a shortage of clerks. "We need typists!" was a common call as first sergeants, and sometimes commanding officers, demanded volunteers. When no one stepped forward, they tended to "check brass" and pull out anyone with AGC insignia and put them to work. Everyone else played basketball. My infantry brass always saved me. And I loved basketball.

It was good to get through the active reserve commitment, but even during the final period of inactive reserve I lived with the prospect of being called up with little advance notice if the Army determined a sudden need for troops. There were a great many areas of international tension, hot spots that might boil over at any time. The threat of getting a call that would put me back in uniform was real, hanging over my head like the proverbial sword of Damocles, but fortunately that call never came.

And yet, had I felt that my service was needed in defense of our nation, I would have gone willingly, as Oscar Koch did. I faced no equivalent of the Great War, no belligerent action like the sinking of the Lusitania, and nothing comparable to World War II, a war our nation had to fight. The Korean War was over and although American involvement in Vietnam already had begun this was still pretty much a national secret.

General Koch was a patriot who loved his country. Not what General Leonard Wood called a "panic patriot," but a thoughtful citizen who accepted the fact that human behavior is such that there always will be threats of war and if war comes it is important the good guys win. World War II would be a classic example. Mass atrocities such as Hitler's murder of millions of Jews and the German, Japanese, and Italian efforts to conquer other nations could not go unchallenged.

Helping the good guys win was Koch's mission as a staff officer. He had abundant confidence Patton would make the right decisions if he had adequate and correct intelligence. He understood very well that, even with good intelligence, winning a battle may come at a terribly high price. I think he would have acknowledged the authenticity of E. B. Sledge's 1981 book, *With the Old Breed*, a compelling account of the U.S. Marines' struggle with the Japanese for control of the Pacific islands of Peleliu and Okinawa.[3]

Robert Hays photo

A robust Gen. Oscar Koch in 1966, a few months before he was diagnosed with terminal cancer.

Photo courtesy Southern Illinois University

The author, at left, with Gen. and Mrs. Oscar Koch and Charlene in 1969. The ravages of the general's illness are apparent.

Gen. Oscar Koch receives a flag for the Carbondale Memorial Day Association's National Flag Bank from Mrs. Burrell Smith. The flag draped the casket of her husband, former Chicago White Sox and Detroit Tigers baseball player James Benjamin "Candy" Smith, a World War I veteran.

Photo courtesy Southern Illinois University

Gen. Koch displays a few of his medals.

Gen. Oscar Koch, at home in Carbondale, Illinois, admires a picture of his old commander, Gen. George S. Patton Jr.

Robert Hays photo

Gen. Koch and Nan display an anniversary plaque received from friends in Austria.

The battle for the tiny island of Peleliu cost the 1st Marine Division more than 6,500 casualties and the Army's 81st Infantry Division more than three thousand. It is estimated that almost eleven thousand Japanese soldiers died defending the island. In combat, Sledge writes, "compassion for the sufferings of others is a burden to those who have it."

Allowing for the accuracy of Sledge's proposition, I expect most men in combat bear the burden of compassion. These lines from one of Rudyard Kipling's most poignant "Departmental Ditties" always have rung true to me:

> The deaths ye died I have watched beside
> And the lives ye led were mine.[4]

General Oscar Koch was a soldier who was well acquainted with the brutality of war. I remember him as a gentle and compassionate man, and find it hard to believe his compassion deserted him on the battlefield. There is a place in war for such men. Soldiers wounded in combat are likely to find compassion in comrades and medics. Squad mates may show it when one of their buddies falls on hard times. I found it even in basic training, where drill sergeants are expected to be masters of the art of verbal abuse at the very least and even physical threat if the occasion demands.

Having barely settled into my new way of life at Fort Leonard Wood and still struggling with military ritual, I pulled my first guard duty on a very cold night. My post was the perimeter of a large, apparently deserted building somewhere in the middle of nowhere. Snow and ice covered the ground and there was an icy wind. Near the end of my second watch—meaning it was the middle of the night—a jeep pulled up and the officer of the guard alighted and approached me slowly. Military protocol called for me to salute, of course. It's just that I wasn't yet all that sharp on military protocol.

The officer, a young lieutenant, stopped in front of me and waited. I stood like a stone statue, not moving, saying nothing.

He cleared his throat. "Now, how is it?" he said quietly.

"It's cold," I said.

He stood for a moment longer, still waiting for my salute. When none was forthcoming, he said, still speaking softly, "Your relief should be here in about twenty minutes. Can you make it that long?"

"Sure. I'll be all right."

He started back to his jeep, then stopped and turned back toward me. "Keep moving," he said. "That will help fight off this cold." Then he drove away.

At about the time his vehicle's taillights disappeared around a corner, I realized what I'd just done. "How is it?" was his gentle way of saying, "Soldier, you're supposed to salute!" Not only did he not chew me out for my blatant failure, but he showed true concern for my welfare, honest compassion. I wish I knew his name. I hope that, if he ever ended up in combat, his compassion did not become a burden.

I am embarrassed to tell about how slow on the uptake I was that night at Fort Leonard Wood, but even more so— now that I know a great deal more about those who fought Oscar Koch's war—admitting I complained about the cold. I wore the 6th Armored Division patch on my shoulder that night and General Robert W. Grow's "Super Sixth" was one of Patton's units that broke through to Bastogne in the Battle of the Bulge. Grow's soldiers saw some of the heaviest fighting in that frigid winter of 1944.

The extent to which 6th Armored Division troops had to contend with the weather as well as the enemy was part of the story. It truly is hard to imagine combat under such conditions:

> Bastogne brought a new experience. Snow, ice and sub–freezing weather provided the setting for one of the most severe campaigns ever fought by American troops. Tank turrets froze, had to be chipped free to regain traversing action. Iced breech blocks had to be manually operated. M–1s refused to function until bolts were beaten back and forth with grenades. When escape hatches and tank doors stuck fast, they got "blow torch" treatment. Ice formed in gas tanks and clogged lines. Feet froze. Men became so cold they "burned."[5]

The winter months at Fort Leonard Wood were very cold, yes, but we weren't in combat. No one was shooting at us. And we never spent many hours in the cold before we were back in heated shelters. I lived in two different barracks and both were warm, heated by coal–fired furnaces which assured the odor of coal smoke was always present.

Even today the aroma of burning coal puts me back at that Missouri military post where my Army service began. I know as well as anyone how fortunate I am my recollections of military service center on things as innocuous as coal smoke and KP duty. I did not have to go to war.

While it may be true that war often brings out the finest personal traits of an individual soldier—traits like courage and honor and self-sacrifice—war itself is a brutal expression of hubris at its worst. Even General Sun Tzu, the ancient Chinese military leader whose renowned treatise, *The Art of War*, has survived the ages, acknowledged, "There is no instance of a country having benefited from prolonged warfare."[6]

Since World War II, we've seen the war in Korea, the long-running Vietnam conflict, the Khmer Rouge regime's genocide in Cambodia, the systematic killing of Muslims by Bosnian Serb forces in the early 1990s, the Rwandan genocide during the same period, and the brutal suppression of opposition in Iran, Libya, and Syria. And of course, the long-running American military activities in Iraq and Afghanistan and Russia's unprovoked attack on Ukraine. In other words, there is persuasive support for Oscar Koch's position. There will be wars that must be fought when inhumane powers, left unchallenged, slaughter innocents. And when wars must be fought, it is important the good guys win. Can we even imagine the world that would have existed had Hitler prevailed?

Oscar Koch had much the same view of war as did his old friend and Patton protégé, General Grow. Grow was said to have hated war, but if there was one he didn't want to be left out.[7] Even Patton, who claimed that battle "is the most magnificent competition in which a human being can indulge," broke into tears at the sight of severely wounded veterans and blamed himself for not being a better commander.[8]

The carnage of battle may be difficult to imagine for

those of us who never have experienced it.

Oscar Koch knew it, and was reluctant to talk about it. I've found this true of most armed forces veterans who have been in combat.

Yet this terribly ugly side of war has been vividly depicted by way of the written word since the earliest examples of Western literature. *The Iliad of Homer* contains gory descriptions of hand–to–hand combat that bring the violence home to the reader. It was Homer, after all, who gave us the familiar phrase, "bit the dust." His stunning accounts of Greek and Trojan soldiers being decapitated by the single slash of a sword are not easy to forget. Homer's epic poetry magnifies the all–encompassing scope of war:

> And as when under the screaming winds
> the whirlstorms bluster
> on the day when the dust lies deepest
> along the pathways
> and the winds in the confusion of dust
> uplift a great cloud,
> such was their indiscriminate battle,
> and their hearts were furious
> to slaughter each other
> with the sharp bronze . . . .[9]

Homer's portrayal of slaughter in a war nearly thirty centuries ago, discomfiting as it is, hardly bests more modern reporting. A graphic *New York Times* story about the 1860 massacre of some two hundred peaceful Native Americans at a settlement near Eureka, California, called Indian Island is but one of hundreds of newspaper entries during the nineteenth century that illustrate the savage treatment of the Native Americans by both white settlers and soldiers who represented individual states or the national government.

The *Times* correspondent found no reason to doubt reports the intruders knew the Indians were without weapons and therefore defenseless and that, "entering lodge after lodge, they dirked the sleeping, and with axes split open and crushed the skulls of the women and children."[10]

While it might be argued this atrocity was not an act of war, a *Times* editorial on September 15, 1865, noted that "Many of the Western settlers are very anxious for a war of extermination against the Indians ...." For all practical purposes, such a war was waged—and it came on the heels of a war to end the indefensible policy of allowing legal slavery in our country.

A good description of combat in the Civil War also streamed from the pen of a journalist, Stephen Crane. His classic novel, *The Red Badge of Courage*, offers a young Union Army soldier's view of the senseless stampede of infantrymen into the teeth of enemy fire: "There was an ominous, clanging overture to the charge when the shafts of the bayonets rattled upon the rifle barrels .... It was a blind and despairing rush by the collection of men in dusty and tattered blue, over a green sward and under a sapphire sky, toward a fence, dimly outlined in smoke, from behind which sputtered the fierce rifles of enemies."[11]

However foolhardy the Civil War tactic of charging across open ground toward an unexposed enemy may have been, the ugly trench warfare of World War I was hardly an improvement. Frank Richards, a British infantryman, describes from personal experience the Battle of Ypres, where enemies faced off at pointblank range. He tells of a firefight which left many German soldiers lying dead in front of the British position and describes how a comrade stood and threw down his weapon, then "jumped on the parapet with his hands above his head" and pointed to a wounded German who was trying to crawl toward them.

In action apparently not all that unusual during what was then known as the Great War, "He then went forward, got hold of the wounded man and carried him in, the enemy clapping their hands and cheering until he had disappeared into our trench."[12]

During World War II, journalist Ernie Pyle wrote eloquently about the common soldier. And like Homer, Pyle also painted the war with a broad brush. He vividly described the "accumulated blur, and the hurting vagueness of being too long in the lines," the total exhaustion, the loss of comrades, and "the constant march into eternity of one's own small quota of chances for survival." Those, he wrote, are among "the things that hurt and destroy."[13]

When it comes to the sad conflict in Vietnam, I need neither literature nor journalism to find a reliable portrait. Letters from two friends who served there kept me abreast of what they experienced as it happened. They managed the highs and the lows of combat with equilibrium and came to take danger in stride.

Lee Hill, a Carbondale neighbor, wrote that while he generally felt safe within his compound, enemy rockets were an ever-present danger. "They come in hissing," he wrote, "and explode with an earth–shaking crack. The best place in the whole world to be then is at the bottom of a hole in the ground, shaking and thinking about dying."[14]

The other man, Mike Lopez, had been a friend since he

was a high school student and I was in charge of the education page of the *Granite City Press–Record*, an Illinois newspaper in the suburban St. Louis area. He was a sensitive kid and a budding photographer who had set out to study photo–journalism at the University of Missouri before changing direction and joining the Army. In a letter written during one of several periods of hospitalization with the exotic fevers common to American soldiers in Vietnam, he wrote, "I can't think of any place I've ever been that is as depressing as an Army hospital. Particularly over here, where the dustoffs are coming in and out. It is kind of sickening and at times seems so useless that so many have given their lives."

In Vietnam, "dustoff" was the radio signal calling for medevac helicopter teams to treat and pick up wounded soldiers.

This young warrant officer was a helicopter pilot. I think he flew medevac missions sometimes, but certainly he flew assault teams into combat. As he neared the end of his tour of duty he was more conscious of the lives he might have taken: "I don't know, maybe I've just been here too long. Got to figuring the other day, and I figure I've accounted for about 100 lives, or at least was a member of a fire team that killed that many people. I guess my major question is, just what have we done? And I guess I'm kind of wondering if it's been worthwhile."[15]

Among anti–war novels, Dalton Trumbo's *Johnny Got His Gun* stands out as one of the most compelling. He intended for his book to make a strong statement, and it does. Trumbo writes about the suffering of a wounded soldier in World War I, describing his injury as a hole that "began at the base of his throat just below where his jaw should be" and extended upward in a widening circle. "He could feel his skin creeping around the rim of the circle. The hole was getting bigger and bigger. It widened out almost to the base of his ears if he had any and then narrowed again. It ended somewhere above what used to be his nose."[16]

Modern warfare rarely involves hand–to–hand fighting or stabbing sleeping women and children, and the "sharp bronze" of Homer's day is no longer in the arsenal. We assume no sane commander would send lines of infantrymen charging across open space straight into the teeth of enemy rifles and cannon. And it is hard to imagine a return to the trench warfare of World War I. But the fact that artillery shells fired from a distance and bombs dropped from high altitude may make the killing impersonal to the killer, battlefield rockets and guided missiles launched from unmanned drones or from ships miles from land strike like a bolt of lightning out of a clear sky—these do not make war less brutal. Battles still lead to the deaths of those who do battle as well as innocents, and battles still are indiscriminate.

Oscar Koch told me, "War is ugly. The only good war is the one you don't fight."

Oscar Koch knew war.

Chapter 22

ON A BEAUTIFUL Sunday morning in June 1969, my friend Tim Turner called me at home and said Old Main, the iconic original building on the campus of Southern Illinois University, was burning. He was in the university's PR office, where we'd worked together before my move, and he was a notorious prankster. I responded with something facetious.

He said, "No, I'm serious. You had better get over to campus." His somber tone told me this was no joke. He said the blaze was close to being out of control, and the firemen were afraid they might not be able to stop its spread to nearby structures. These would include Susan B. Anthony Hall, where both Turner's office and mine were located.

"Do you think we may have to start moving stuff out?" I asked.

"They'd like us to be ready, just in case."

Authorities knew from the beginning the fire, which destroyed the magnificent sandstone and red–brick building that had stood as a symbol of the university since 1886, was a work of arson. Separate blazes had been started in both the north and south ends of the attic, which housed an ROTC firing range, and in stairwells below. A vulgar message had been scrawled on a chalkboard to call attention to the deed.

Turner and I watched from a third-floor Anthony Hall window as the firemen struggled to contain the blaze to the single massive structure. It was three hours or more before we felt comfortable our building was safe and late afternoon before firemen said the fire was under control. The battle to extinguish it went on for several hours more.

Although the arson was an act of outrageous malice, no one raised questions about the arsonist's motive. Protests against continuation of the war in Vietnam had grown both more common and more forceful, especially on college campuses around the country. They would get worse as the months moved by, making for another year of contentious social disruptions.

The year had begun with Richard Nixon assuming the office of president. In a later address to the nation, he would call on the "silent majority" of American citizens to support his policies in Vietnam. Students at Harvard University took over University Hall, demanding an end to the ROTC program on that campus. The Weathermen, a protest organization, took control of the Students for a Democratic Society national office, and antiwar protests grew even stronger in the fall when Lieutenant William Calley was charged with six counts of premeditated murder for his role in the 1968 My Lai massacre in Vietnam. Before the year ended, the nation would see its first draft lottery since World War II.

The antiwar movement caused both honest concern and a certain level of national paranoia. Columnist Paul Scott reported in October that teams of riot watchers, headed by Lieutenant General William J. McCaffrey and operating out of war room–like basement quarters in the Pentagon, was on alert for expected nationwide demonstrations. He said they were analyzing FBI and local police reports on the activities and movements of "two dozen known pro–communist militants highly placed in the antiwar groups."[1]

Scott went on to say the "most disturbing development uncovered so far by the federal riot watchers is the organized effort by the antiwar groups and their professional agitators to recruit high school and grade school students in Washington and other cities to join in the antiwar protests."

Despite the growing antiwar sentiment, though, not all the important events of 1969 were negative. The year went down in history primarily because men landed on the moon. The July 20 landing by NASA's Apollo 11 spacecraft and images of Commander Neil Armstrong stepping onto the moon's surface caught and held the attention of the world.

A great many lesser happenings merit mention. *Slaughterhouse–Five*, the still–popular antiwar novel by Kurt Vonnegut Jr., and Mario Puzo's *The Godfather* were published and "Sesame Street" made its debut on PBS television, along with network shows like "The Brady Bunch" and "Rowan & Martin's Laugh–In." The musical group Led Zeppelin released its first album, while the Beatles performed their final public concert and Creedence Clearwater Revival released "Bad Moon Rising," one of its greatest hits. The movie, "Midnight Cowboy," was released in May, "Easy Rider" in mid–July, and then "Butch Cassidy and the Sundance Kid" in September.

Other events that stood out even in the incredible year of 1969 were the Woodstock music festival in upstate New York and the murders of actress Sharon Tate and others in the home of Hollywood figure Roman Polanski by Charles Manson and his "family" of followers. Both events were in August.

And Murray Gell–Mann won the Nobel Prize in physics for his work on the theory of elementary particles and discovery of the quark. As a non–scientist, I usually wouldn't have paid all that much attention to a Nobel Prize in physics. But Gell–Mann was the brother of my friend Ben Gelman, a talented photographer and columnist on the Carbondale newspaper, *The Southern Illinoisan.*

Ben would be instrumental in bringing his brother to campus as keynote speaker for a conference of scientists and, partly because of Ben, I used a report on Murray's visit as the cover story in the next issue of the university's *Alumnus* magazine. And if the topic comes up in conversation, I likely will be the only one who knows what a quark is.[2]

A few days before the moon landing, the *New York Times* had run a tongue–in–cheek correction of a 1920 "Topics of the Times" editorial page feature that dismissed the possibilities of space flight because, it contended, a rocket couldn't function in a vacuum. But, the *Times* now admitted, "Further investigation and experimentation have confirmed the findings of Isaac Newton in the 17th Century and it is now definitely established that a rocket can function in a vacuum as well as in an atmosphere."[3] Humor, even in small doses, was appreciated.

On our campus, although it would be some months before the anti–war protests reached their riotous peak, the arsonist had inflicted an enormous wound. Where Old

Main had stood like a proud citadel in the center of the original campus there now was a gaping hole. A great deal of classroom space had been lost. Also left in the rubble were the university museum and the main offices of several academic departments and, along with them, a number of graduate students' on–going research projects and theses and dissertations in progress.

The loss of Old Main had hurt much more than mere institutional pride. I would write in *Alumnus* magazine: "She was a magnificent old queen, dominating a campus and symbolizing a University. Her loss is not the death of the institution, not even a lasting cripple. But it is a scar, deep and painful and permanent."[4]

In spite of the daily distractions in the world around us and the general's weakened physical condition, though, in August we finished the manuscript for *G–2: Intelligence for Patton*. It was a thin volume. There were topics I would have liked to cover that we never got around to, steps along the way during the general's decades of loyal service to his country I would have liked to treat more fully. But what had been my worst fear—that General Koch might not live to sign off on the completed work—came to naught.

I suspect if he had been stronger, able to devote more time to his effort, he would have had suggestions that under the circumstances were not voiced. He made notes on his usual yellow pad, commenting on a line, a paragraph here and there. Nothing dramatic. No errors of fact came to my attention. To the very end I was impressed with the general's writing skills, his ability to rephrase a sentence. After all, that was my job. I was the city editor Robert S. Allen had sold him on.

And to the end, also, General Koch's modesty prevailed.

I can cite passages in the book where he should claim more credit. Sometimes I'd tried to write things that way, only to face objections. It was his book; it was written his way.

The general sent a copy of the manuscript to Patton's son, Colonel George S. Patton III, and invited his comments and especially "any corrections on the chapter titled 'George Smith Patton, Jr.'" He also sent a copy of the magazine article I'd done on "the so-called intelligence failure attending the Battle of the Bulge." He said he regretted "none of the authors of books on the Bulge—John Eisenhower included—ever took the time to do the research on this vital issue."[5]

General Koch mentioned John Eisenhower because Eisenhower's book on the Battle of the Bulge, *The Bitter Woods*, had been published just recently.

Robert Allen had written the general earlier, "I haven't read all of his [Eisenhower's] book, but the portions I have read left me very cold and scornful. Not only does he ignore Intelligence, but he also is equally sketchy about the crucial role that Patton and Third Army played in overcoming this disaster."[6]

The younger Patton had written of his interest in our book a few months earlier, and noted the Vietnam war "has really put intelligence in the forefront where it should be." And he added, "As you may know, it's called the G–2/S–2 war by many."

After reviewing the manuscript, Colonel Patton wrote that he found it most interesting and timely "in view of the tremendous effort now being placed on our intell. activities. You can't fight the VN type war without intelligence." If the book didn't get published, he said, he'd like a copy of the manuscript. And if it did make it into print, "I'll buy several."[7]

For the general and me, completion of the manuscript meant reaching our primary goal. The collaboration had run on for three years. When I thought back to the beginning, I recalled the Memorial Day observance in Carbondale's Woodlawn Cemetery, the centennial celebration of the 1866 ceremony at which General John A. Logan was the featured speaker. General Koch had called my attention to the closing words of Logan's address the first time we talked about plans for the Woodlawn observance.

Although General Logan's address that day was not recorded, the cemetery caretaker, James Green, jotted down a few notes on the speaker's remarks in the margin of a Bible. According to Green, who happened to be Logan's cousin, General Logan ended his speech with the words, "Every man's life belongs to his country and no man has a right to refuse it when his country calls."

General Koch found Logan's words to be quite moving. He chose them as a theme of the printed program for the centennial celebration. He said they were as relevant in 1966 as they had been a hundred years earlier.

The general had arranged for color guards representing all the military services. They came from the 101st Airborne Division, Fort Campbell, Kentucky; the Second Coast Guard District headquarters, St. Louis; the Memphis Naval Air Reserve Training Center; and the Military Air Lift Command headquarters at Scott Air Force Base, Illinois. The Air Force also sent a bugler and a ceremonial firing squad.

Thinking back to that day with the perspective gained since, I understood more clearly why this had been a labor of love for Oscar Koch. He had taken General Logan's words to heart because they were words he had lived by.

His country had called and he had served, willingly and well.

Decades later, he still had a clear recollection of the day he took the first step on his own long and eventful road as a soldier. He wrote in a memo marking the fiftieth anniversary of his enlistment in the Light Horse Squadron Association in Milwaukee in 1915, "I raised my right hand and solemnly swore to defend these United States against all enemies, whomsoever."

General Koch left no doubt his determination to serve was prompted in large part by patriotism, and he was prepared to fight for his country: "It was on that date that I thought I was doing my share, making my contribution to fend against the infringements of our American rights, which the sinking of the Lusitania was forecasting."[8]

I hoped the general's love for his country, as I had come to see it, would be apparent in *G–2: Intelligence for Patton.* Surely his dedication would. And his contributions.

First, though, we had to clear the next hurdle.

We had no control over this one, and quite likely it would be higher. We had to find a publisher.

General Koch's physical condition left little hope he would be able to involve himself in this undertaking. I had only the skimpiest notion how to go about it. I'd been writing free-lance article proposals for magazines for several years, though, and it seemed logical similar queries to book publishers would be an acceptable approach to this new challenge. With antiwar sentiment at its height, publishers hardly would be eager to bring out books relating to anything military—and things were not getting any better.

May 1970 brought antiwar protests home to Southern Illinois University. The month began with the tragedy at

Kent State University, where four students were killed and nine wounded when National Guard troops opened fire on student protestors. The shooting came after hours of tense confrontation. James A. Michner, in his detailed and fascinating account of what happened at Kent State, reports "at least two and probably seven students" were bayoneted by guardsmen the night before. None was seriously injured, though two were hospitalized.[9]

After the Kent State shootings and escalation of the war by the U.S. invasion of Cambodia, serious protests began on my campus. The student senate voted unanimously for an indefinite boycott of classes and Chancellor Robert MacVicar canceled classes for a period of mourning honoring the Kent State students.

Things deteriorated from there. Thousands of students took to the streets and marched downtown, closing intersections and trying to block the Illinois Central railroad tracks. Illinois State Police were called out in force to backstop campus and city officers and more than five thousand Illinois National Guard troops were mobilized and stationed in Carbondale. Many of the guardsmen were students themselves and generally sympathized with the protesters.[10]

After there had been unsuccessful attempts to firebomb the campus ROTC headquarters, Wheeler Hall, many university faculty and staff members, including me, volunteered to serve as "fire guards." We worked in pairs, spending nights in our respective buildings and alternating stints of sleeping and walking the hallways.

Police cars and National Guard jeeps patrolled the streets after both the city and the university declared curfews, and the odor of teargas permeated the community. As demonstrations and confrontations continued and the situation grew more dangerous, Chancellor MacVicar announced the university would close with the end of classes on Friday, May 15. There would be no commencement exercises and no summer session. More than four hundred arrests had been made and scores of campus and downtown buildings had been damaged. Damage, for the most part, was confined to broken windows.

General Oscar Koch would hardly be aware of these things. His life was slipping away.

As I recall his final days, I'm struck by the emotional courage he exhibited throughout his long battle with cancer. He was a man who enjoyed life. He no doubt would have been happy to walk this earth for years longer had he been granted that time. But he accepted death with grace, suffered without complaint.

My last conversation with the general took place in a cramped patient's room at the U.S. Veterans Hospital in Marion, Illinois. We both knew the end was near. His spirit was not diminished. Nan was there. She had brought him a small block of cheese and he began to nibble at it as soon as she managed to free it from a foil wrapping.

"He's like a mouse when it comes to cheese," she said, patting his hand lovingly.

The general, all color drained from his face now, smiled his best smile. We clasped hands and, without using the words, I told him goodbye.

Chapter 23

AS IF IN THE blink of an eye after our final handshake, I found myself standing at the front of a small chapel in the Huffman Funeral Home in Carbondale about to deliver Oscar Koch's eulogy. Nan had asked me to. She said there was no one more appropriate, that the general had come to look on me much as if I were a son. I was honored and humbled. I had been given but four short years with this man and he had made my life better in more ways than I could count. As deeply as I felt his loss, I felt equally blessed for having known him.

I knew this was going to be difficult. I wrote into a prepared script a couple of early end–points to give myself a way out if I became too emotional. I read the eulogy to Nan and explained I might have to stop speaking at one of these places if I simply wasn't able to go on.

"Don't be embarrassed if you have a problem," Nan said. "I saw Oscar almost break down giving the eulogy for John Allen." I tried to remember why I had to miss John Allen's funeral, but I couldn't recall.

I stood at the front of the room, close to the general's

flag–draped coffin. I was accustomed to having Mary somewhere nearby to lend her support at times like this. But she had just come down with a virulent case of the flu and couldn't be there. I understood how hard this was for her, because she loved Oscar Koch and she would have been there had it been at all possible.

I looked over the room and noted a number of friends among those gathered to pay their last respects to the general. One of them was my former boss, Bill Lyons, head of the Southern Illinois University news bureau. He nodded encouragement.

The last person to arrive was Bill O'Brien, a former college football coach best known as a National Football League referee. He was something of a celebrity in the community and we frequently saw him on television on Sunday afternoons during the football season when he called games involving our favored St. Louis Cardinals. But I had forgotten he was a colonel in the U.S. Marine Corps Reserve. I looked up to see who entered, and saw him in his dress uniform. I knew he wore it to honor General Koch. The lump in my throat expanded and I felt tears welling up in my eyes.

There was some preliminary word and then it was my time to speak. I began:

> Yesterday, May 16, 1970, was Armed Forces Day—a day set aside by the American people to honor those who serve our country in the uniforms of the various services. It is appropriate, perhaps, that on this day there departed from us a gallant soldier, Brigadier General Oscar W. Koch, United States Army.[1]

I told of the general's remarkable military record in some detail. I came to the first "stopping point" written into my comments, and summoned the courage to keep going.

"General Koch was too modest to dwell on his own successes," I said. "But he took pride in doing to the best of his ability any task assigned. It was this dedication to duty which enabled him to forecast in the bitter winter of 1944 the coming enemy surge in what history would record as the Battle of the Bulge. And it was this same dedication which led General Patton to assure the highest Allied commanders, concerned about a critical mission assigned to a staff officer, 'Don't worry, he can do it; he can do anything.'"

My voice faltered. I had to pause for an instant before I continued:

> After World War II, General Koch organized and directed the first peace-time combat intelligence school in the history of the United States Army. He served as director of intelligence for the high commissioner and commanding general of U.S. forces in Austria. And when this nation went to war again, this time in Korea, he served there as assistant commander and then commander of the historic American combat unit, the 25th Infantry Division.

I talked about General Koch's contributions to the Carbondale community. It seemed to me even many of those gathered here probably were not aware of all the things he had done. With this, I managed to get past the final "stopping point." And I'm glad I did. At the conclusion I wanted to add my final personal observations.

"Yet to his family and friends—those who really knew him—these are not the things Oscar Koch will be remembered for," I said. "He will be remembered for his marvelous spirit, his wit, his twinkling eye. He will be remembered for his concern for others, his willingness always to lend a helping hand. He will be remembered for his gentle manner, his love of life and his ability to make the best of every day. He will be remembered for his courage, his courtesy—and for his ready laughter. He will be remembered for all those qualities, large and small, which mark a good man.

"He once wrote of his old commander and comrade that GI's still stand a little straighter when they tell you, 'I served with Patton.' I contend that all of us stand a little taller just for having known Oscar Koch."

I noted that it had been a quarter–century since the end of the war in Europe. I paid tribute to the many thousands of soldiers to whom we owed so much and who must be counted as the true heroes of our age. I said one of the foremost among them was General Oscar Koch.

"We salute you, Sir," I concluded. "We commend you on duties well done. And we bid you Godspeed."

After the services in Carbondale, Oscar Koch's body was taken to Arlington National Cemetery for burial. There, on a rolling green hillside not far from the eternal flame marking the grave of President John F. Kennedy and even closer to the burial site of U.S. Supreme Court Justice Oliver Wendell Holmes, the general lies at rest. The peaceful setting is in marked contrast to the battlefields where he so ably served his country. Yet it is a site that seems exquisitely appropriate. Around it, in every direction, lie the graves of other soldiers, a few marked by splendid monuments but most displaying only simple white stones perfectly aligned in row upon row.

Among the many who expressed their condolences to Nan was Colonel George S. Patton III. He wrote that Oscar Koch "was always true to all of us, especially daddy. We of the Patton family will treasure his memory."[2]

Although I know it was cancer which took the general's life, his death certificate listed the cause of death as "hardening of the arteries." Nan showed it to me, with a sense of accomplishment. "See," she said, "it doesn't say 'cancer.'" I can only assume she'd appealed to an understanding doctor not to mention the disease she so despised.

Nan Koch lived on to age ninety–six. She suffered Alzheimer's disease in her declining years and passed away on March 12, 1995, at the Abbey of Carbondale nursing home.

She enjoyed one final appearance in the limelight. Soon after *G–2: Intelligence for Patton* was published, the touring Royal Lipizzan Stallions show was booked at the Southern Illinois University Arena. Nan invited Mary and me to attend with her, as her guests. We were barely seated when someone from the Arena management came and asked me to verify that Mrs. Oscar Koch was there, and checked her seat number.

At intermission, the public address announcer reminded the audience General George S. Patton Jr. was credited with saving the Lipizzans from slaughter in Czechoslovakia in 1945.[3] Then he said the management was proud Mrs. Nan Koch, the widow of Patton's G–2, Oscar Koch, was among those present. He asked her to stand and be recognized. She did, and suddenly she was brightly spotlighted, the sole focus of a brilliant beam from high in the center of the Arena ceiling.

Others in the huge audience stood and applauded. I saw that Nan had tears in her eyes, but her pride was evident. She had devoted most of her adult life to Oscar Koch, and this was for him. Oscar would have been thrilled to see her have her own turn in the public eye.

I felt a surge of pride. I was happy for Nan, and I was happy that it had been my privilege to know and work with Oscar Koch. We had finished his book, and I was more determined than ever to see it in print. It was time the world heard about his enormous contribution to the fight for freedom from tyranny that had been our national cause.

Chapter 24

WITH THE GENERAL'S passing, I was left to face the daunting challenge of finding a publisher for his book, *G–2: Intelligence for Patton*. Alone. This proved to be every bit the trial I had expected, and then some. Most of the publishers I contacted rejected the project out of hand. A few asked to see the manuscript, then turned it down. It seemed books about war and the military were not particularly popular at this time.

I was encouraged when a university press asked to see the manuscript. This particular press had published a number of works in military history.

My hopes were very high as I put the manuscript in the mail. After six months with no decision, I sent a letter asking for the return of my material. The press director wrote back promptly, apologized for the delay, and asked if I might give him a bit more time. The press was very much interested in *G–2*, he said, but he had not been able to retrieve the manuscript from an "expert reader" whose approval was needed before a decision could be made on publication.

My response to his response was an easy one.

General Oscar Koch was the foremost authority on the subject. He would not need another "expert" to judge his work. Please return the manuscript as soon as possible. In retrospect, I probably should have been more patient. Had *G–2* been solely my work, I most likely would have given up. A drawer of my filing cabinet was filling up with rejection letters, most of them highly positive toward the subject but declaring the book did not fit the respective publisher's needs at this time. Prospects for finding a publisher had come to look pretty hopeless. But I had promised General Koch I'd see the book through to publication, and there was no way I could in good conscience cease to try. Not until I had exhausted every possibility I could find. I would continue my search.

And then a bit of luck. I was aware of the 1967 book, *Warrior: The Story of General George S. Patton, Jr.* I picked up a copy from the library and gave it a closer look. It was written by the editors of *Army Times* and, more important, it was published by Putnam's as "An *Army Times* Book." Might the *Army Times*, which was published in Washington, D.C., and billed itself as the "national weekly newspaper for the United States Army," be receptive to lending its name to *G–2: Intelligence for Patton*?

I sent a hurried query.

There was a prompt response from Adolph A. Hoehling, The Army Times Publishing Company book editor. First of all, I did not need to sell them on the importance of Oscar Koch. They knew very well who he was and they knew his record. And yes, they had published books under the *Army Times* imprint—but they no longer did.

My heart sank as I read those last words. Had I missed perhaps the last opportunity to get *G–2* into print? But then I read on.

Hoehling believed General Koch's book surely merited publication. He would like to see the manuscript. He could make no promise, but there was at least an outside chance he could gain support for one more book. And he emphasized once more he could not promise a positive result.

I sent him the manuscript and waited nervously for further developments. In short order he reported back that, although he still faced some internal opposition, The Army Times Publishing Company would publish the general's book. I never knew how much opposition Hoehling had to overcome inside the company, but I applaud him for standing up for a project he believed in.[1]

G–2: Intelligence for Patton was published in 1971 by Whitmore Publishing Company of Philadelphia as an *Army Times* Book. Not only did it contain less meat than it might have had it not been for General Koch's illness, but it also lacked maps and photos, and even an index, all sacrificed in an effort to keep the price low. Difficult as it is to imagine today, the goal was to keep the selling price no higher than five dollars. It sold for $4.95. Yet time would prove it had much to offer.[2]

I was proud when *G–2: Intelligence for Patton* was published, and did what I could to help promote it. Newspaper coverage was good. I did interviews on a few radio stations and made a limited number of television appearances. At a Missouri CBS station a young news anchorman rushed in, unprepared, and began to ask about my association with "General Oscar Cook." It was a recording session and since we weren't being broadcast live, he stopped and started over after I corrected him.

I was the sole guest on an hour–long talk show on an

Illinois ABC station. The host began with a good-natured joke: "You're not going to use any of Patton's colorful language today, are you?"

All the broadcast journalists were pleasant and competent professionals, but my memories of these sessions are mostly unpleasant. I found it difficult to be sitting in a guest chair that should have been occupied by Oscar Koch. I sensed that, after a time, Nan Koch also saw this as unfair. But when I asked a television host to limit the emphasis on my role in the book and stress General Koch as the author, he told me, rather forcefully, "General Koch is gone. You have to be his voice now."

Not surprising, given the antiwar atmosphere of the time, there was not an immediate groundswell of interest in *G-2* among the reading public. The book got good reviews in the popular, mainstream press and military journals such as *Infantry* and *Armor*, however.[3] And one library journal review, which I consider to be among the most thoughtful, concluded *G-2* was "an excellent book, particularly for the non-military." The reviewer contended the book would "help dispel many of the misconceptions concerning this service" and "show that the persons performing it are sensitive and sincere human beings."[4]

From the outset, there was little doubt in my mind *G-2* eventually would take root among military historians. However, I was aware of studies showing it normally takes several years for newly published information to make its way into other books.

In the case of *G-2*, though, it happened sooner than I expected. General Hubert Essame's thorough 1974 biography, *Patton: a Study in Command*, draws heavily on our book and labels Oscar Koch the "spark plug" of Third Army.

Because Essame chose not to use source citations, it is not apparent to his readers that several pages and a number of shorter passages in his work are taken almost verbatim from *G–2: Intelligence for Patton*. I was not offended by this. *G–2* is included in his extensive bibliography, and it was good finally to see General Koch get the attention he deserved.

Essame writes that Patton had in Oscar Koch, "what was probably, in the field of intelligence, the most penetrating brain in the American Army." He calls Koch "the most brilliant and original member of Patton's command team."[5] And unlike others who focus principally on Koch's role in the period preceding the Battle of the Bulge, Essame bases his evaluation on Koch's entire period of service with Patton.

Nan Koch presented me with a copy of Essame's book shortly after it became available. She inscribed it, "To Robert G. Hays, in appreciation for helping to make *G–2 Intelligence for Patton possible.*"

Over time, *G–2* began to surface more frequently as a reference that counterbalanced the prevailing "intelligence failure" view of the Battle of the Bulge. The record on that count had become a good deal more clear by the early 1990s, with *G–2* having become a widely used source of information. A number of new books on World War II in Europe, and particularly accounts of the Battle of the Bulge, had more fully acknowledged the contributions of Oscar Koch.

Charles B. MacDonald, in his epic 1985 book on that battle, *A Time for Trumpets*, reports in detail on the December 9, 1944, briefing in which Koch spelled out to Patton the seriousness with which he viewed the German buildup. This was the meeting at which Patton, following his intelligence chief's report, vowed to be ready for "whatever happens."[6]

Geoffrey Perret's excellent book, *There's a War to be Won*, published in 1991, generously credits Oscar Koch's intelligence work as the basis for Patton's preparations in advance of the Bulge. Meanwhile, Perret writes, intelligence chiefs in the higher headquarters—those of Generals Bradley and Eisenhower—had "handsful of dust that they threw in each other's eyes."[7] (This remains one of my favorite descriptions of the failures at that level in the weeks preceding the Bulge.)

Charles M. Province's 1992 book, *Patton's Third Army*, notes that a Koch intelligence report on the concentration of enemy forces in the Ardennes area was sent to Eisenhower's headquarters as early as December 9, 1944, but "was ignored." Then, on December 16, Province writes, "The repeated warnings to SHAEF by Patton's G–2 officer, Colonel Koch, were finally borne out."[8]

Carlo D'Este's masterpiece, *Patton: a Genius for War*, published in 1995, gives Oscar Koch virtually full measure of the credit due. D'Este, like Essame, includes Koch's contributions over the duration of World War II. D'Este makes note of the fact that Oscar Koch was among the very small number of men who formed Patton's "inner circle," and that he'd been a friend and associate of Patton since well before the beginning of the war. He also pays high tribute to Koch's outstanding performance as Third Army G–2, particularly in the period preceding the Battle of the Bulge. He says, "Koch was the only Allied intelligence officer to anticipate trouble and plan how to deal with it. Thus, where other intelligence officers were lulling their commanders with false optimism and wishful thinking that nothing serious was imminent, the Third Army made plans to deal with what no one else believed would occur."[9]

George Forty's *The Armies of George S. Patton*, published in 1996, says in a chronology of Third Army's battles that "On 9 December Third Army G–2 sent a report to SHAEF indicating a probable enemy offensive in the Ardennes, but it was ignored." Further, when the German offensive was launched, "Thanks to Colonel Oscar Koch's insistence that the Germans would launch an all–out offensive in the Ardennes and Patton's eminently sensible orders to plan for every emergency, Third Army was 'ready to roll' whenever it was given the green light."[10]

In his intricately detailed biography of Patton, Stanley P. Hirshson says of Koch that from November 20 on he "reported the growing strength of the Sixth Panzer Army and warned of a German counteroffensive."[11]

Tributes, regardless of the intent, sometimes turn out to be a bit flawed. One such appears in recorded testimony before the U.S. Senate Select Committee on Intelligence in July 2004. William E. Odom, who had a distinguished military career and served as director of the National Security Agency in President Ronald Reagan's second term, offered a comparison of the relationship between senior commanders in World War II and their intelligence officers.

Odom contended there were, indeed, "several bits of intelligence" suggesting the Germans were preparing for a counteroffensive in the period before the Battle of the Bulge. He noted that Montgomery and his G–2 "stubbornly rejected the facts; Bradley and his G–2 remained skeptical and passive." And Eisenhower and his intelligence officer "were somewhat quicker to sense the danger but slower than Patton, whose G–2 saw it coming several weeks beforehand, prompting Patton to initiate contingency plans to respond to it."

Odom cited an interesting study by Harold C. Deutsch, who claimed the dominating personalities of the commanders likely intimidated their intelligence officers. But not Patton. Odom noted Deutsch's conclusion that, "Perhaps the fine performance of Gustave Koch [Patton's G–2] was largely due to being lucky in his boss."[12]

Gustave? Oscar probably would not have complained.

Unfortunately, there still are more recent instances in which good military historians have failed to recognize the work of Oscar Koch and his Third Army G–2 Section. Stephen Ambrose, who writes about the buildup to the Battle of the Bulge from the perspective of both the Allies and the Germans in his 1997 book, *Citizen Soldiers*, shows little awareness of the intelligence reports from Third Army. He cites the 21st Army Group's intelligence summary of December 16, stating the enemy "is at present fighting a defensive campaign on all fronts; his situation is such that he cannot stage major offensive operations."[13] He notes that neither Eisenhower nor Bradley believed there was any danger of a German attack.

Ambrose writes that Bradley was still skeptical at the end of the day, dismissing the enemy advance as being "of little consequence" even as reports of the German action came in, while Eisenhower accepted the obvious and recognized it as a counteroffensive. He says of the December 19 meeting of the Allied leaders at which Patton let it be known he was ready to launch an attack on the German forces from the south, "Eisenhower's lieutenants entered the room glum, depressed, embarrassed, as they should have been, given the magnitude of the intelligence failure ...."14

Without explanation, Ambrose notes that before Patton left Third Army headquarters for this meeting, "he had told his staff to begin the preparations for switching his attack line from east to north." He gives Patton credit for his "boldness" in being prepared and Eisenhower credit for his "decisiveness" in responding to the German attack. He does not mention Oscar Koch.

Even more remarkably, though, there still is serious misinformation afloat. As late as 2005, John S. D. Eisenhower wrote in *The Patton Saber*, a publication of the Patton Museum Foundation at Fort Knox, Kentucky:

> When Adolph Hitler launched his gigantic counteroffensive in mid-December, 1944, Lieutenant General George S. Patton, Jr., was as surprised as everyone else, from the Supreme Commander on down. Patton, lacking the intelligence resources available at higher headquarters, was slower to recognize the offensive for what it was: an effort to break the Anglo–American lines through the lightly–defended First Army position in the "impassable" Ardennes with the object of seizing the major port of Antwerp 15

Eisenhower apparently intended for this article to be complimentary to General Patton, claiming Third Army was highly effective in responding to the German breakout in spite of not having advance intelligence. Factually, however, he missed his target by miles and miles.

While I fervently wish General Koch had lived to see the success of *G–2: Intelligence for Patton* in terms of setting the record straight and giving him credit for his brilliant work, I believe he would be most pleased with the role the book has played in strengthening intelligence training in the nation's armed forces. It has, for all practical purposes, become the text book the general had in mind long before I became involved. It is on a variety of required reading lists throughout the military services, and I learned it was used as a text in the intelligence training program at Fort Drum, New York, when a supply officer there called after the original edition was out of print and asked if it might be reissued. I was uncertain at the time, but it was not long afterward when the Schiffer Publishing Company brought out a new paperback edition.

Most satisfying to General Koch would be the cadre papers and student research reports at the U.S. Army War College, Air University, and elsewhere in the service schools that draw on the book as a solid launch point for better intelligence training. He would see each of these as a tribute to his firm conviction that intelligence officers are made, not born.

Lieutenant Colonel Michael D. Rosenbaum, in a 1999 War College strategy research project report, wrote that future intelligence operations should not be driven by technology but, rather, by "well–trained intelligence professionals, who have studied history and understand doctrine and the intelligence battlefield operating system." Rosenbaum drew heavily on *G–2*, which he described as "a superb book," and also used Koch's original intelligence reports archived at the Military History Institute in Carlisle, Pennsylvania.[16]

In his research, Rosenbaum studied Oscar Koch's Third Army intelligence operations, principally during the period September through December 1944, leading up to the Battle of the Bulge. He argued Koch's work provided lessons that "are cornerstones upon which the professional intelligence officer can depend in a volatile, uncertain, ambiguous, and complex world."[17]

Rosenbaum found lessons in Oscar Koch's pre–Bulge intelligence activities that, he said, "provide the basis for some of our intelligence doctrine today and that can serve as cornerstones for intelligence doctrine and operations in the Army After Next." Further, he found that Koch's work demonstrated the need for well–trained intelligence professionals. "When one looks at the Third Army intelligence reports, with the knowledge of what actually occurred," he declared, "it is easy to see the prophetic nature of COL Koch's estimate."[18]

In what surely would have brought a broad smile to General Koch's face, Rosenbaum called for a new appreciation of the old doctrine that the intelligence officer's responsibility is to determine the enemy's capabilities rather than his intentions.

Noting that the Army's 1994 field manual on "Intelligence and Electronic Warfare Operations" calls for predictive intelligence, he contended that predictive intelligence must be based on an underlying assumption "that a key decision already has been made" by the enemy. He said prediction is always accompanied by a certain degree of risk and argued that the degree of risk acceptable is properly a command decision.

Further, Rosenbaum cited a 1945 examination of military intelligence operations ordered by the Secretary of

War which concluded that expecting the G–2 to offer the commander "conclusions as to enemy intentions rather than on presenting facts bearing on enemy situation and capabilities" was a misunderstanding of the proper function of intelligence. The report said this transferred an important command responsibility from the commander to his G–2.[19]

General Oscar Koch would have concurred fully with Rosenbaum's position. It further vindicated his persuasive argument about the role of intelligence in combat.

In Koch's eyes, the role of intelligence—his role as G–2—was simple. It was to help his commander make good decisions. Patton would have concurred, and both he and Oscar Koch would have acknowledged that the process traveled a two–way street.

Today there is a Military Intelligence Hall of Fame at Fort Huachuca, Arizona. Oscar Koch, of course, is among those honored with induction. It seems immensely fitting that his supporting biographical information is titled, "Oscar Koch and the Confidence of the Commander." Surely no team in the history of the American armed forces offers a better example of the relationship between a commander and his G–2 as it ought to be.

Chapter 25

WHEN THE MOVIE, "Patton," hit the screens of the nation's theaters in 1970, movie critics were almost unanimous in proclaiming it was destined to become a classic. The movie includes a scene in which General George S. Patton Jr., as portrayed by actor George C. Scott, takes his staff by surprise when he predicts German forces are about to launch an Ardennes counteroffensive: "There's absolutely no reason for us to assume that the Germans are mounting a major offensive. The weather is awful and their supplies are low. The German Army hasn't mounted a winter attack since Frederick the Great. Therefore I believe that's exactly what they're going to do."

Among those who appear to be most surprised by this pronouncement is Colonel Gaston Bell, the movie version of Patton's G–2, a role ably played by actor Lawrence Dobkin. Colonel Bell apparently can't imagine such a scenario as his commander has just projected.

Colonel Gaston Bell is the fictitious character created by the movie makers when Oscar Koch declined to approve the script provided by producer Frank McCarthy. The movie is admirable in many ways despite its flaws and

George C. Scott's performance is perhaps the premier example of superb screen acting during my lifetime, but there was no Colonel Gaston Bell and Patton did not make momentous command decisions based on hunches. Such distortions of fact do a great disservice to history.

Had those responsible for the movie been concerned about accuracy, countless movie-goers through the years might have learned it was Oscar Koch who, through brilliant intelligence work, predicted the German counteroffensive that launched the Battle of the Bulge. A true image of Koch might have emerged, and today his name might be widely recognized among those of the great American military figures of World War II.

The character Colonel Gaston Bell was no more a figment of a screenwriter's imagination than the totally erroneous impression of Patton's relationship with his G-2 which is left by the movie. I believe the movie exaggerates the competitiveness between Patton and Montgomery in Sicily and goes to extremes to make Bradley look good, perhaps reflecting the fact he served as a technical director.[1] There also are other significant factual errors that are bothersome, but the total erasure of Oscar Koch and the incalculable impact he had on Patton's success are to me the most egregious.

Oscar Koch didn't live to see the movie, and given the way it distorted his relationship with Patton I take some comfort in that fact. Surely he would have felt it a degrading experience. And although I think he would have admired the heroic characterization of his old commander, he would have felt there was an injustice in portraying Patton as an audacious risk-taker. As Conrad C. Crane has observed, Patton was aggressive and willing to take advantage of opportunities, but worked diligently to mitigate risks. He gave intelligence a high priority and "Oscar Koch always had first say in planning."2

Nan Koch saw the movie and was highly critical.

She felt it did not do justice to Patton, to say nothing of its flawed portrayal of his G–2. She also was not impressed with George C. Scott's depiction of Patton. She admired Patton very much and John Wayne, she said, would have been a better choice for the role: "He's a cowboy, horseback–riding type of person. He's the only actor who could have closely carried out the part."3

Not only did both General Koch and Nan love Patton, but they also held Patton's widow, Beatrice, in high regard. "She never forgot a member of the Old Man's staff," the general said affectionately.

One of his most prized mementos of a long and colorful military career was the Patton desk clock that sat on the mantle of his home in Carbondale. He received the clock in April 1946 from Beatrice. She enclosed a note which read, "Dear Col. Koch, Georgie always used this clock on his office desk before he went to war. I gave it to him, so, in a way, it is from us both—with deep appreciation."

I understand the general's gratitude because I have a similar treasure. A cherry–wood valet stand that traveled with Oscar Koch across continents and met his needs through World War II and beyond stands now at the foot of my bed. Nan Koch gave it to me after her husband's death because she wanted me to have, in her words, "something personal, something Oscar used every day."

I'm living my own retirement now, far removed from the young man who proudly wore the uniform of the U.S.

Army, the young man who stood in awe of Oscar Koch simply because he was a retired general. I'm still proud I wore the uniform and I still view Oscar Koch with a sense of wonder, but now because I know his remarkable record of service to his country.

I am grateful I never had to go into combat. I am grateful my sons never had to go to war. I hope my grandson doesn't, that no one else's sons and grandsons do. I understand the horrors of war. I know World War II, Oscar Koch's war, took the lives of perhaps sixty million men, women, and children around the globe and brought incalculable suffering to millions more.

I do not believe General Koch found glory in war and I am quite certain that, even if he did, he sought none for himself. I believe, also, that he did not view himself as a commander, or leader of soldiers in battle. His post as 25th Infantry Division commander in Korea was brief. I never asked him about this period in his career and he never brought it up.

He was a staff officer. A superb one. And while he was highly conscious that his work as G–2 affected the commander's decisions in vital ways directly related to the fate of troops in combat, he tended to think in terms of lives saved rather than lives lost. This might be viewed as meaningless semantics, but in his mind there was an important distinction. He often made the point that the impact of intelligence errors would be measured in terms of lives lost. The obvious flip side of this tenet is that good intelligence results in lives saved.

And Oscar Koch practiced good intelligence.

While most of us go through life and barely leave a few ripples, Oscar Koch left gigantic waves for which we all are indebted. In his unassuming, craftsman–like manner, he went about his duties determined to do the best he could. His best proved to be remarkable. How do you quantify contributions that saved perhaps thousands of American soldiers' lives, helped release Europe from Hitler's mad grasp, and helped preserve freedom where it existed in the rest of the world?

The general's competence as an intelligence officer is a matter of record. There is more. Like Ladislas Farago, who writes that Oscar Koch deserved the Medal of Honor for standing up for his convictions even in the face of strong opposition from higher authority, Major Kevin Dougherty cites Koch's unwavering moral courage as one of the principal reasons for his success.

Dougherty is one of many to label Koch an unsung hero behind Patton's victories. Considering Koch's performance before the Battle of the Bulge, Dougherty writes, "When he began reporting that the German Army was far from destroyed and even capable of counterattacking, his was clearly a voice in the wilderness." And at the same time, while Koch was urging caution, Patton was planning a Third Army offensive and initially was skeptical of his G–2's findings.

"To offer a view that contradicted his higher headquarters and at the same time brought him into opposition with as strong a personality as Patton's required great moral courage" by Koch, Dougherty writes. Further, while Patton is remembered as one of the war's great heroes, Koch remains relatively unknown. "Such is the nature of behind–the–scenes staff work."[4]

It is time for General Oscar Koch's name forever to be included on the roster of World War II heroes, unsung no more. Anything less flies in the face of recorded history.

Today we are separated from those difficult and painful war years by the decades that have come and gone, yet I find it easy to slip back in memory to that time. But I was a child then, and my reminiscences are more pleasing.

On my desk today sits a beautiful and perfectly scaled nine–inch, die–cast model P-47 Thunderbolt fighter plane. One glance at it and I'm a boy again, back in Walnut Grove school. I twirl the propeller with my finger. I can hear the roar of its high–powered Pratt & Whitney engine. We run outside. I imagine that Nina, Donnie, Ann, and Marilyn are there beside me, looking to the skies, fascinated by those exquisite machines of war.

Recollections of my months in training at Fort Leonard Wood and my inauspicious service as a regimental special orders clerk at Fort Jackson still are vivid. My easy time with the physical rigors of basic training was in stark contrast to the apparent ordeal these posed for older men in my training company. I remember a man we called Pinkie, who had served in the Army before, reached the rank of sergeant, and then become a civilian. Training was hard for Pinkie. He was in poor physical shape after civilian life. A recruiter had promised he wouldn't have to go through basic training again, but once he was back in uniform the Army took a different view.

Pinkie's last military assignment of his earlier tour of duty had included riding "shotgun" once a month on an armored car bringing the Fort Leonard Wood payroll from St. Louis. The Army paid in cash, and Pinkie told about sitting on millions of dollars, holding a machine gun across his lap and agonizing over the fact he was guarding a treasure while all he had to look forward to was the trifling salary of an Army enlisted man.

At the time I volunteered, the draft age had crept up to 27 as quotas were lowered with the end of the war in Korea. Many of those drafted with me were men who had finished college, established themselves in professions, begun families. Military service had not been in their plans. Surely they missed their families at home. Surely they were lonely.

All soldiers probably experience loneliness at one time or another. I remember driving through St. Louis on my way to Fort Leonard Wood, listening to The Platters' recording of "The Great Pretender" on my car radio and feeling I was all by myself in an empty space between the civilian world I'd left and the military world I was barely into. And I remember, too, driving past a small house alongside U.S. Route 41 somewhere in Tennessee on my way to Fort Jackson. It was twilight and lights glowed inside, reminding me a family was there, probably sitting down to supper, while I had miles to go and hours of solitude to endure.

I feel selfish now complaining of loneliness during my undemanding period of military service. Soldiers like Oscar Koch and many of the young draftees called to war in the 1940s and 1950s were there for the duration. Many never would see home again.

Not only did I avoid such hardship, but I've lived a long and full life. I have a wonderful family and keeping in touch with former students helps me not to feel old. I'm proud of their successes and pleased when they let me know they are going to be back in town and would like to get together. Eddie, the extraordinarily handsome and intelligent orange tabby cat that belongs next door, drops by almost every day and lets me know I'm still on his list of favorite humans. He forgives my faults and I enjoy his company. I'm comfortable in our modest home, which I can maintain without too much effort, and our surroundings are pleasant. We have dogwood trees and redbud and a tulip magnolia which burst forth in the spring to signal new life, followed by the weigela and mock orange. The oaks and hard maple bring glorious color in the fall.

Last year I brought home a pair of mimosa seedlings from South Carolina. Mimosas sprout up like weeds there, and many South Carolinians consider them a nuisance, but central Illinois winters are too severe for mimosa to thrive. I know of only two of them in town. I planted mine in sheltered places where they would be protected from icy north winds and both survived.

I look at them and wonder if the mimosa tree I planted in Carbondale long ago is still alive.

Southern Illinois winters are much kinder than those we experience here. Mimosa and crape myrtle are not exotic there. The tree in question is visible in one of my favorite pictures of Mary, sitting on a step in front of our house on Walkup Street. It is the one that dropped its final blossoms the day I first heard the distressing news about General Koch's cancer.

Cancer is a word none of us wants to hear. I heard it a few years ago, but prompt surgery at the skilled hands of Dr. Richard Wolf seems to have taken care of mine. I am well beyond that point where I can safely call myself a cancer survivor. If the general's illness had arisen today, perhaps he could have been saved. I don't know. But I wish he could have been spared the suffering.

My diagnosis led me to retire from teaching, a move

that was past due. Since then I've had more time to devote to writing. I'm working on a fifth novel. Maybe I'll add a character based on Oscar Koch. He could be a soldier or a priest. A senator. A university president. Perhaps merely a kindly grandfather. He could be any of these and be authentic, unreconstructed from the man I knew.

Sometimes when I've been sitting at the computer too long and need a break from the keyboard, I switch from word processing to the music library conserved on the machine's hard drive. Il Divo, probably, or Andrea Bocelli, or maybe the old Neil Diamond classics. Or I may go to YouTube on the Internet and watch musical videos. A current favorite is a 2005 Pink Floyd reunion performance. I see the band members, no longer the young men we want to remember, and marvel that their talent is undiminished. I'm drawn in easily by their classic execution of "Comfortably Numb," and I am fascinated by David Gilmour's magical ability to wring more angst from a guitar than should be humanly possible.

I watch the group in other concerts and I notice fan reaction is the same wherever they perform. Music is universal. I wonder why we can't take some of the money spent on war and use it to help bridge the world's cultural gaps with music.

Not an original thought, of course. Ernie Pyle, who was closer to the American soldiers in World War II than any other journalist, writes that he heard soldiers say a thousand times, "If only we could have created all this energy for something good."[5]

On cool autumn evenings we sit before the fireplace, Mary and I, and listen to the music of Bobby Hackett.

"Mood Indigo" begins and I'm instantly transported

back to Fort Leonard Wood in early 1956. I lie on my bunk while another man retrieves a small record player from his locker. Outside the night is terribly cold. Inside, on the second floor, the barracks is warm and comfortable. In administrative school there is no demanding physical activity.

Today we learned how to update and maintain the thick book of Army regulations. Tonight we will fall asleep to the soothing tones of Bobby Hackett's trumpet.

Just now I watch through a dining room window as squirrels ravish a buckeye tree in the back yard. This is a peaceful setting, the natural world as it should be, the world as it has existed for millennia, withstanding the assaults of mankind. I easily forget the wars of the past, the wars ongoing now elsewhere on our fragile planet.

Today there is something new, something that wasn't there yesterday. On the next street over, barely visible above the roof of a house in between, stands a new flagpole. The pole is white, topped by a brass–colored fitting, a globe, or perhaps an eagle, which glistens in the brilliant afternoon sunlight. I can barely see the top edge of an American flag raised to the full height of the pole as it waves in a strong southwest breeze.

I think back to the Koch home in Carbondale, where the general proudly flew this flag every day. I'm grateful for Nan's invitation to that special Saturday morning flag-raising, grateful I was there to see the general's obvious pleasure in teaching the young Boy Scouts the proper handling of this banner he had honored for a lifetime. I'm grateful I was witness to the boys' excitement at being the center of his attention and grateful for their demonstration that they, too, will honor this flag.

I think back to Oscar Koch as I knew him.

I mention his name to Mary. "I still miss him," she says, and her eyes mist. I still miss him too.

And I choose not to remember Oscar Koch in death. Instead, I will remember him in life—vital, strong of character, always giving the best of himself.

General Koch was a Rotarian. Not long after we'd begun work on *G–2: Intelligence for Patton* he was called on to be the featured speaker at the Carbondale Rotary Club's weekly luncheon. His topic? The importance of intelligence in the Patton commands. He invited me to be his guest and I enthusiastically accepted.

Then he said, "I want you there for support. You sit over on one side and every few minutes nod your head yes, like you agree with everything I say."

"But general," I protested softly, "how could I possibly add anything? You're the world's foremost expert on this topic. You don't need anyone's support."

"I don't care," he said. "We're a team and intelligence is a team activity. That's the way we did it in Patton's briefings. No matter who made the presentation, the others sat alongside and nodded support. That's exactly what I need you to do, if you don't mind."

General Koch made a superb presentation, of course, and I dutifully sat at the side of the room, frequently nodding my head. I felt somewhat awkward and hoped nobody noticed. But the general noticed.

"Thank you," he said after the meeting. "You made me feel like I was back in the G–2 tent reporting to Patton. That's what teamwork is all about."

He smiled that infectious Koch smile. He was pleased. I was pleased, as well. I had just been on the receiving end of a modest—but to me, immensely gratifying—example of command support. And it came from Brigadier General Oscar Koch, Patton's G–2, a man who always will be my hero.

The End

Appendix

Following is the eulogy to General Oscar Koch delivered by Robert Hays at the general's memorial service in Carbondale, Illinois, on May 17, 1970.

Yesterday, May 16, 1970, was Armed Forces Day—a day set aside by the American people to honor those who serve our country in the uniforms of the various services. It is appropriate, perhaps, that on this day there departed from us a gallant soldier: Brigadier General Oscar W. Koch, United States Army.

Oscar Koch devoted a lifetime to the service of his country. That service began more than a half-century ago, on June 18, 1915, when as a mere lad of 18 he raised his right hand and solemnly swore to defend these United States against all enemies. And we can imagine the pride of a young man as he left the armory that evening, a member of what was then called the "Light Horse Squadron Association." That was in Milwaukee, Wisconsin, where Oscar Koch was born on January 10, 1897.

The Association was an exclusive outfit. Members paid dues toward maintenance of the armory and equipment. A private was paid $15 a month and a lance corporal was a man to be reckoned with. And an 18-year-old couldn't join without parental consent. This apparently was a painful decision for young Oscar's widowed mother. But the argument which finally won her over was his own simple statement that he probably wouldn't see any action, anyway.

A year later, however, the Association—now designated as Troop A, First Wisconsin Cavalry—was on duty with General John J. Pershing on the Mexican border in pursuit of the elusive Pancho Villa. The direction of Oscar Koch's life had been set. He was a soldier. He would pass through every existing rank from private to general officer. And, despite his earlier assurances to his mother, in nearly four decades of service he would see action in three major wars and make his mark as one of the most brilliant intelligence officers in American military history.

In the vanguard of American troops to arrive in Europe in World War I, he was commissioned a second lieutenant on his twenty-first birthday. He became an instructor in the famed French artillery school at Saumur, and in action across France he first saw some of the places with which he would become much more familiar a war later.

Returning to the states, he organized and commanded the first federally-recognized National Guard unit in the state of Wisconsin—and this at the age of 22! In 1920 he resigned his Guard captaincy to accept a commission in the Regular Army Cavalry. In those early years of peace he was to make important contributions as an instructor in the Army Cavalry School. And, as recognized in the official records of the Army Signal Corps, he developed the first air-to-ground pickup device.

It also was during this period that Oscar Koch and some other young Cavalry officers took their mounts to perform in a horse show at the Iowa State Fair. There he met and instantly became attracted to a young woman from Carbondale, Illinois, Miss Nannie Caldwell. They were married on August 2, 1924.

From that day on, Nan Koch has lived those moments that only a military wife can know. She survives her husband, along with his two sisters, Mrs. Lillian Larson and Mrs. Marjorie Gleason, and a nephew, Carbys Gleason, all of California.

When the United States entered World War II, Oscar Koch was called to combat by one of America's great military leaders, General George S. Patton Jr. He had served under Patton in time of peace; when Patton went to war he wanted men of proven ability at his side. Thus Oscar was to serve as chief of staff for Taskforce Blackstone in the invasion of French Morocco and then as Patton's chief intelligence officer for the remainder of the war—with the II Corps and I Armored Corps in North Africa, the Seventh Army in its conquest of Sicily, and finally with the Third Army in its operations across Europe.

General Koch was too modest to dwell on his own successes. But he took pride in doing to the best of his ability any task assigned. It was this dedication to duty which enabled him to forecast in the bitter winter of 1944 the coming enemy surge in what history would record as the Battle of the Bulge. And it was this same dedication which led General Patton to assure the highest Allied commanders, concerned about a critical mission assigned to a staff officer, "Don't worry, he can do it; he can do anything."

After World War II, General Koch organized and directed the first peace-time combat intelligence school in the history of the United States Army. He served as director of intelligence for the high commissioner and commanding general of U.S. forces in Austria. And when this nation went to war again, this time in Korea, he served there as assistant commander and then commander of the historic American combat unit, the 25th Infantry Division.

We would be greatly remiss, of course, if we cited Oscar Koch only for his military career. His contributions were too many and too varied for that. After his retirement from the Army in 1954, he received a grant from the Guggenheim Foundation to support research in military history, the first military man ever to receive such recognition from that respected institution.

Certainly his contributions to the Carbondale community deserve more than passing attention. He spearheaded the Carbondale Memorial Day Centennial Committee's drive for national recognition. His efforts led to establishment of the National Flag Bank. He served as a director of the Jackson County YMCA. He was a Mason and a member of the Elks Lodge. And I'm sure his fellow members of the Carbondale Rotary Club could give a long list of his contributions to that organization. Oscar Koch was called upon frequently because he was a man who could get things done.

Yet to his family and friends—those who really knew him—these are not the things Oscar Koch will be remembered for.

He will be remembered for his marvelous spirit, his wit, his twinkling eye. He will be remembered for his concern for others, his willingness always to lend a helping hand.

He will be remembered for his gentle manner, his love of life and his ability to make the best of every day. He will be remembered for his courage, his courtesy—and for his ready laughter. He will be remembered for all those qualities, large and small, which mark a good man.

He once wrote of his old commander and comrade that GI's still stand a little straighter when they tell you, "I served with Patton." I contend that all of us stand a little taller just for having known Oscar Koch.

During the past winter we witnessed the 25th anniversary of the Battle of the Bulge. Just recently we were reminded that it has been a quarter-century since the end of the war in Europe.

Thousands of men across our nation looked back at the dark days of that conflict and re-lived in memory that part of their lives. Some wondered, perhaps, if what they accomplished was worth the sacrifices made. We know, of course, that it was. It is to these men that we owe our freedom and even our very lives. They must be counted among the real heroes of our age. One of the foremost among them was General Oscar Koch.

We salute you, Sir. We commend you on duties well done. And we bid you Godspeed.

Endnotes

Chapter 1

1. Carol Alexander, "Patton Film 'did not do him justice,'" *The Southern Illinoisan* (Carbondale, Ill.) July 10, 1970. Although the author had the letter in hand on the day it arrived, the text of the letter quoted here is from this news report.

2. Brig. Gen. Oscar W. Koch with Robert G. Hays, *G–2: Intelligence for Patton*, an Army Times Publishing Company Book (Philadelphia: Whitmore Publishing Co., 1971).

3. "Again—A Warning Against Whipping up War Hysteria," *Capital Times*, Madison, Wis., April 22, 1966.

4. Logan originally had been a Southern sympathizer, but before the Civil War began he became a strong supporter of Abraham Lincoln and the Union. For a brief but comprehensive biography, see "John A. Logan" in John W. Allen, *Legends and Lore of Southern Illinois* (Carbondale: Southern Illinois University Area Services Division, 1963), 28–30.

5. Program, National Memorial Day Centennial, Carbondale, Illinois, May 30, 1966 (Springfield: Division of Tourism, Illinois Dept. of Business and Economic Development).

Chapter 2

1. Judge Charles Kerr, ed., *History of Kentucky*, 5 (Chicago: The American Historical Society, 1922), 619. John Caldwell died in office. For a comprehensive account of George Rogers Clark's strong influence on Kentucky polities, see Kerr, 1, 325–35.

2. Isaac Caldwell had taken a summer job with the Illinois Central Railroad which he enjoyed so much he did not want to leave it for West Point. He became a conductor and stayed

with the railroad for more than 50 years.

3. Gen. Omar N. Bradley, *A Soldier's Story* (New York: Henry Holt Co., 1951), 33.

4. H. Essame, *Patton: A Study in Command* (New York: Charles Scribner's Sons, 1974), 24.

5. Kim Hollen, "Patton Warned of Pearl Harbor Attack," U.S. Army News Service, (Arlington, Va.), Dec. 6, 2010.

6. Robert G. Hays, "He Helped Decide to Hold Bastogne," *St. Louis Post–Dispatch* "Everyday Magazine," August 30, 1966.

Chapter 3

1. Stanley P. Hirshson, *General Patton: A Soldier's Life*, Perennial ed. (New York: HarperCollins, 2003), 469.

2. Col. Ralph Luman to Nan Koch, May 22, 1970. Author's files.

3. Dr. Murray Zimmerman to Robert Hays, Jan. 28, 2000. Author's files.

4. Cyrus R. Shockey, "Memories of General Grow and General Patton," 6th Armored Division web site, posted Nov. 17, 1998.

5. Robert S. Allen, *Lucky Forward* (New York: The Vanguard Press, 1947), 49–50.

Chapter 4

1. Brig. Gen. Oscar W. Koch with Robert G. Hays, *G–2: Intelligence for Patton* Schiffer Military History ed. (Atglen, Pa.: Schiffer Publishing Ltd., 1999).

2. Allen, *Lucky Forward*, 49.

3. Hirshson, *General Patton*, 683.

4. Oscar Koch, *Aide Mémorie*, Mar. 16, 1960. Author's files. Gen. Koch's careful attention to detail led him to keep extensive

notes in the form of memoranda such as the one cited here.

5. *Ibid.*

6. Henry Allen Moe to Oscar Koch, Jan. 4, 1962. Author's files.

7. James McGovern to Oscar Koch, July 6, 1962. Author's files.

8. Oscar Koch, *Aide Mémorie*, nd. Author's files.

9. *Ibid.*

10. Ladislas Farago to Oscar Koch, Jan. 18, 1963. Author's files.

11. Oscar Koch, *Aide Mémorie*, nd. Author's files.

12. Ivan Obolensky to Oscar Koch, July 12, 1963. Author's files.

13. Ladislas Farago, *Patton: Ordeal and Triumph* (New York: Ivan Obolensky, Inc., 1964), 573–74.

14. Oscar Koch to Anon. April 2, 1964. Copy in author's files.

15. Anon. to Oscar Koch, April 4, 1964. Author's files.

Chapter 5

1. The author is grateful to Maj. Ray L. Hays (U.S. Army, ret.) not related but a former colleague at the University of Illinois, for a photocopy of the original document of surrender. Maj. Hays was on the staff of Gen. Dwight Eisenhower and was present at the signing.

2. To read more about Walnut Grove basketball, see Robert Hays, "At Walnut Grove the uniforms almost fit," *WomenSports*, 3 no. 4 (April 1976), 50–51.

3. Capt. Ted W. Lawson, *Thirty Seconds over Tokyo* (New York: Random House, 1943).

4. Robert Hays, *Blood on the Roses* (Everett, Wash.: Vanilla Heart Publishing, 2011), 175–76.

Chapter 6

1. Irving Dilliard, foreword to John W. Allen, *Legends and Lore of Southern Illinois,* v.

2. Robert Hays, "Teller of Tales," *SmallWorld,* 6 no. 3 (Summer 1967), 8–9.

3. Oscar Koch, "John Willis Allen, October 14, 1887–August 29, 1969," Sept. 1, 1969. Delivered at John W. Allen's memorial service.

4. Koch (with Hays), *G–2: Intelligence for Patton,* 33.

5. *Ibid.,* 82.

6. *Ibid.,* 33.

Chapter 7

1. Phillip Harkins, *Blackburn's Headhunters* (New York: W.W. Norton, 1955).

2. Gen. Blackburn's biography on the Arlington National Cemetery web site notes that "much of what he did as a senior Army official and strategist remains top secret. He was in command of covert unconventional warfare in Southeast Asia during the Vietnam War to undermine the spread of communism, and later spent several years at the Pentagon and in other military roles in the Washington, D.C., area."

3. The base scuttlebutt over time indicated that Gen. Costello actually was highly regarded by the Fort Jackson senior officers.

4. "Remembering Our Local Italian Americans, Past and Present," web site of the Columbia, S.C., Lodge of The Sons of Italy in America, accessed Feb. 4, 2012.

5. Allen, *Lucky Forward,* 51, 52.

Chapter 8

1. Joe W. Wilson, The 761st "Black Panther" Tank Battalion in World War II (Jefferson, N.C.: McFarland & Co., 1999), 3.

2. Hirshson, *General Patton*, 524.

3. Ray Sprigle, "A Soldier Who Came Home to Die," *Pittsburgh Post–Gazette*, August 9, 1948.

4. Dave Hinton, "New movie offers reason to recall Tuskegee Airmen's time in Rantoul," *The News–Gazette* (Champaign, Ill.) Jan. 22, 2012

Chapter 9

1. Dr. Murray Zimmerman to Robert Hays, Jan. 14, 2000. Author's files.

2. Louis Cassels, "Largest Antiwar Rally in Capital Held Peacefully," *The Sunday Herald* (Provo, Utah), Oct. 22, 1967.

3. Richard Breitman, "Historical Analysis of 20 Name Files from CIA Records," research report, Interagency Working Group, American University, April, 2001. National Archives, declassified records of the Central Intelligence Agency (RG 263).

4. Seventh U.S. Army, *The Seventh Army in Sicily*, Part 2. Sec. C, c–6 (Palermo, Sicily: Hq Seventh Army, Sept. 15, 1943). Author's personal copy of the Seventh Army Sicilian campaign after–battle report.

5. F. W. Winterbotham, *The ULTRA Secret* (New York: Harper and Row, 1974).

6. Farago, *Patton: Ordeal and Triumph*, 697.

Chapter 10

1. Allen, *Lucky Forward*, 68.

2. Seventh U.S. Army, *The Seventh Army in Sicily*, Part 2, Staff Reports, Sec. C, "Report of the A. C. of S. G–2," c–2, c–3.

3. *Ibid.*, subsection xii, "Conclusions," c–7.

4. Koch (with Hays), *G–2: Intelligence for Patton*, 44. For a detailed overview, see Erasmus H. Kloman, *Assignment Algiers: With the OSS in the Mediterranean Theater* (Annapolis, Md.: U.S. Naval Institute Press, 2005).

5. Michael E. Bigelow, "Big Business: Intelligence in Patton's Third Army," in James P. Finley, ed., *U.S. Army Military Intelligence History: A Sourcebook* (Fort Huachuca, Ariz.: U.S. Army Intelligence Center & Fort Huachuca, 1995), 211.

6. Seventh U.S. Army, *The Seventh Army in Sicily*, Part 1, "Summary of Operations," a–2.

7. Allen, *Lucky Forward*, 51n.

8. Seventh U.S. Army, *The Seventh Army in Sicily*, Part 1, "Summary of Operations," a–2.

9. Ernie Pyle, *Brave Men* (New York: Henry Holt and Co., 1944), 274–75.

10. Seventh U.S. Army, *The Seventh Army in Sicily*, Part 2, Staff Reports, Sec. C, "Report of the A. C. of S. G–2," c–13.

11. *Ibid.*, c–14.

12. Koch (with Hays), *G–2: Intelligence for Patton*, 54.

Chapter 11

1. Charles M. Province, Unknown Patton (On–line, CMP Publications, 2002), 51.

2. *Ibid.*, 57, 32

3. Koch (with Hays), *G–2: Intelligence for Patton*, 157.

4. Dr. Murray Zimmerman to Robert Hays, Jan. 14, 2000. Author's files.

5. *Ibid.*

6. Seventh U.S. Army, *The Seventh Army in Sicily*, Part 1, Summary of Operations, Sec. D, "Planning Instructions No. 1," d–40.

7. *Ibid.*, Part 1, Sec. C, "Lessons Learned, Tactical: mental attitude," c–8.

8. *Ibid.*, Part 1, Sec. B, "The Plan," b–25.

9. *Ibid.*, "Summary of Operations," b–4.

Chapter 12

1. Seventh U.S. Army, *The Seventh Army in Sicily*, Part 1, Sec. B, "Summary of Operations," b–8.

2. *Ibid.*, Part 2, Sec. C, c–33

3. *Ibid.*

4. *Ibid.*, c–35.

5. *Ibid.*

6. *Ibid.*, Sec. B, "Summary of Operations," b–22.

7. *Ibid.*

8. *Ibid.*, b–25.

Chapter 13

1. The general's note to me was, on the whole, pleasant and upbeat. It was clear from what he wrote that in usual Oscar Koch fashion he was determined to make the best of the situation. But any mention of being uncomfortable was a sure sign that he had been in significant pain.

2. Alexander, 'Patton Film 'did not do him justice.'"

3. For a comprehensive study of Koch's use of ULTRA, see Maj. Bradford J. Shwedo, "XIX Tactical Air Command and ULTRA: Patton's Force Enhancers in the 1944 Campaign in France," Air University Cadre Paper No. 10 (Maxwell Air Force Base, Ala.: Air University Press, 2001).

4. Shwedo, "XIX Tactical Air Command," 18.

5. "General Patton and COMINT [Communications Intelligence]," based upon a contemporary report by Maj. Warrick Wallace, Cryptologic Almanac online, U.S. National Security Agency/Central Security Service, posted Jan.15, 2009.

6. Shwedo, "XIX Tactical Air Command," 54.

7. Warrick Wallace, "Report on Assignment with the Third US Army 15 August–18 September 1944," National Archives RG 457, May 21, 1943, cited by Shwedo, 55.

8. "General Patton and COMINT."

9. Bigelow, "Big Business," 212.

Chapter 14

1. Koch (with Hays), *G–2: Intelligence for Patton*, 70.

2. Allen, *Lucky Forward*, 101–108.

3. Shwedo, "XIX Tactical Air Command," 65–66.

4. Koch (with Hays), *G–2: Intelligence for Patton*, 79.

Chapter 15

1. See Koch (with Hays), *G–2: Intelligence for Patton*, 80.

2. For the full story of why Paris was spared, see Larry Collins and Dominique La Pierre, *Is Paris Burning?* (New York: Simon and Schuster, 1965).

3. Maj. Jeffrey W. Decker, "Logistics and Patton's Third Army: Lessons for Today's Logisticians," Air & Space Power Journal, Chronicles Online Journal, Mar. 20, 2003, 11.

4. *Ibid.*, 12

5. Allen, *Lucky Forward*, 132.

6. Decker, "Logistics," 12.

7. Allen, *Lucky Forward*, 136.

8. Decker, "Logistics," 13.

9. Koch (with Hays), *G–2: Intelligence for Patton*, 81.

10. Allen, *Lucky Forward*, 158.

11. Shwedo, "XIX Tactical Air Command," 116.

12. *Ibid.*, 120.

13. Allen, *Lucky Forward*, 30.

14. Shwedo, "XIX Tactical Air Command," 122.

15. Koch (with Hays), *G–2: Intelligence for Patton*, 87.

Chapter 16

1. "A Stupid Attempt to Find a Scapegoat," **San** *Antonio Express*, Dec. 29, 1944.

2. Dwight D. Eisenhower, *Crusade in Europe*, paperback ed. (Baltimore: The Johns Hopkins University Press, 1997), 338.

3. Winston Churchill, *Triumph and Tragedy*

   (Boston: Houghton Mifflin, 1953) 6, 272.

4. Hansen W. Baldwin, *Battles lost and Won: Great Campaigns of World War II* (Old Saybrook, Conn.: Konecky & Konecky, 1966), 316, 357.

5. Hugh M. Cole, "The Ardennes: Battle of the Bulge," United

States Army in World War II, The European Theater of Operations (Washington, D.C.: Center for Military History, U.S. Army, 1993), 63.

6. B. H. Liddell Hart, *The German Generals Talk* (New York: Berkley Publishing Corp., 1958), 226.

7. Editors of The Army Times, *Warrior: The Story of General George S. Patton Jr.* (New York: G. P. Putnam's Sons, 1967), 160.

8. Koch (with Hays), *G–2: Intelligence for Patton,* 97.

9. Eisenhower, *Crusade in Europe,* 351–52.

10. Gen. Omar N. Bradley and Clay Blair, *A General's Life* (New York: Simon and Schuster, 1983), 353–54.

11. D. A. Lande, *I Was with Patton* (St. Paul, Minn.: MBI Publishing Co., 2002), 206–07.

12. Cited by Lt. Col. Michael D. Rosenbaum, "The Battle of the Bulge: Intelligence Lessons for the Army after Next," Strategy Research Project, U.S. Army War College, 1999, 12.

13. Bradley, *A General's Life,* 351.

14. "Oscar Koch and the Confidence of the Commander," Masters of the Intelligence Art, inscription, Military Intelligence Hall of Fame, Fort Huachuca, Ariz.

15. Stephen E. Ambrose, *Citizen Soldiers* (New York: Simon & Schuster, 1997), 184.

Chapter 17

1. See, for example, Robert G. Hays, "The Real Miracle of Bastogne: Gen. Patton's Prayer," Sunday Look, *The Sunday Courier and Press Magazine* (Evansville, Ind.), Dec. 10, 1967; Robert Hays, "The Battle of the Bulge—and Patton's Prayer," *The News—Gazette* (Champaign, Ill.), Dec. 11, 2011; Robert

G. Hays, "The Miracle of Bastogne," *The Link* (Washington, D. C.: The General Commission on Chaplains and Armed Forces Personnel, Oct. 1969), 30–33.

2. Chaplain James H. O'Neill, "The True Story of the Patton Prayer," *The Military Chaplain*, October–November 1948, 1–3.

3. *Ibid.*, 2

4. *Rolling Together*, 4th Armored Division Assn. newsletter, Christmas issue 1965, 3–4.

5. David Maxey to Robert Hays, June 19, 1968. Author's files.

6. John Fink to Robert Hays, Oct. 9, 1968. Author's files.

7. Edward Kern to Robert Hays, July 8, 1968. Author's files.

8. Oscar Koch to Robert S. Allen, Oct. 28, 1967. Copy in author's files.

Chapter 18

1. Koch (with Hays), *G–2: Intelligence for Patton*, 123–131.

2. *Ibid.*

3. Lande, *I Was with Patton*, 224–25.

4. Allen, *Lucky Forward*, 39.

5. Robert S. Allen to Oscar Koch, Oct. 9, 1968. Author's files.

6. Koch (with Hays), *G–2: Intelligence for Patton*, 145.

7. *Ibid.*, 151.

8. Bigelow, "Big Business," 209.

9. Carlo D'Este to Robert Hays, Apr. 23, 1996. Author's files.

Chapter 19

1. Brian Sullivan, "Negro Mobs Stage Rampages in Cities," *The Daily Mail* (Charleston, W. Va.), Apr. 5, 1968.

2. "Wallace Names LeMay No. 2," *Atlanta Constitution*, Oct. 4, 1968.

3. Koch (with Hays), *G–2: Intelligence for Patton*, 115.

4. Bigelow, "Big Business," 211.

5. Koch (with Hays), *G–2: Intelligence for Patton*, 119.

6. Allen, *Lucky Forward*, 49.

7. James P. Finley, "Winning Smart," A Brief History of U.S. Army Intelligence (Fort Huachuca, Ariz., nd), 33.

8. Oscar Koch to anon., Dec. 14, 1968. Copy in author's files.

9. *Ibid.*

10. Seventh U.S. Army, *The Seventh Army in Sicily*, Part 2, "Report of the A. C. of S. G–4," e–37.

Chapter 20

1. Kevin C. Ruffner, ed., "Forging an Intelligence Partnership: CIA and the Origins of the BND, a Document History." CIA History Staff, Center for the Study of Intelligence, European Directorate of Operations, 1, 1999 (declassified 2002), ix.

2. Dr. Murray Zimmerman to Robert Hays, May 14, 2001. Author's files.

3. Glen B. Infield, *Skorzeny: Hitler's Commando* (New York: St. Martin's Press, 1981), 4.

4. Otto Skorzeny, *Skorzeny's Special Missions* (London: Robert Hale Ltd., 1957), 145–46.

5. *Ibid.*, 148–150.

6. Infield, *Skorzeny: Hitler's Commando*, 142.

7. Koch (with Hays), *G–2: Intelligence for Patton*, 157.

Chapter 21

1. Oscar Koch's fellowship application to the John Simon Guggenheim Memorial Foundation, Nov. 9, 1954. Copy in author's files.

2. Robert Hays, *The Life and Death of Lizzie Morris* (Everett, Wash.: Vanilla Heart Publishing, 2008), 119.

3. E. B. Sledge, *With the Old Breed At Peleliu and Okinawa* (New York: Presidio Press, 2007), 158.

4. Kipling's "Departmental Ditties," first published in the late 1800s, have been cited by writers ever since as quick and colorful images from military life. These lines were used as a thematic epigraph in Sledge, *With the Old Breed.*

5. "Brest to Bastogne, The Story of the 6th Armored Division," G.I. Stories series, Stars and Stripes, nd (circa 1945). Link at Super Sixth: The story of Patton's 6th Armored Division in WW II web site, accessed Apr. 10, 2012.

6. Sun Tzu, *The Art of War*, Lionel Giles translation, 1910, Chapter 6, "Waging War," Rule 6.

7. Cyrus R. Shockey, "Memories of General Grow and General Patton," 6th Armored Division web site, accessed May 1, 2012.

8. Province, *The Unknown Patton*, 52; Lande, *I Was with Patton*, 268–69.

9. *The Iliad of Homer*, translated and with an introduction by Richmond Lattimore (Chicago: University of Chicago Press, 1951), 280.

10. Robert G, Hays, *A Race at Bay* (Carbondale: Southern Illinois University Press, 1997), 2.

11. Stephen Crane, *The Red Badge of Courage* (New York: Collier Books, 1962), 119.

12. Frank Richards, "The Battle of Ypres," in Ernest Hemingway, ed., *Men at War* (New York: Bramwell House, 1979), 870.

13. Pyle, *Brave Men*, 451

14. Lee Hill to Robert Hays, Feb. 4, 1968. Author's files.

15. Michael Lopez to Robert Hays, Aug. 3, 1970. Author's files.

16. Dalton Trumbo, *Johnny Got His Gun* (New York: Bantam Books, 1967), 62.

Chapter 22

1. Paul Scott, "U.S. Officials on Alert for Anti–war Violence," *Post-Tribune* (Jefferson City, Mo.), Oct. 13, 1969.

2. While Murray Gell–Mann retained the original spelling of the family name, Ben changed the spelling to Gelman. Ben was nine years older than Murray and also had an abiding interest in science, particularly as applied to the natural environment.

3. "A Correction," *New York Times*, July 17, 1969.

4. Robert Hays, "A Landmark Falls," *Alumnus/Southern Illinois University* (Carbondale, Ill.), 31 no. 1 (July 1969), 10–12.

5. Oscar Koch to Col. George S. Patton III, Aug. 12, 1969. Copy in author's files.

6. Robert S. Allen to Oscar Koch, Mar. 1, 1969. Author's files.

7. Col. George S. Patton III to Oscar Koch, July 15, 1969. Author's files.

8. Oscar Koch memorandum, nd. Copy in author's files.

9. James A. Michener: *Kent State: What Happened and Why* (Greenwich, Conn.: Facwett Publications, 1971), 250.

10. Stephanie Moletti, "Spring 1970: a Season of Protests," *Daily Egyptian* (Carbondale, Ill.), Special Report, nd, 1995.

Chapter 23

1. Robert Hays, "Oscar W. Koch, Jan. 10, 1897–May 16, 1970." Eulogy delivered at Oscar Koch memorial service, Carbondale, Ill., May 17, 1970.

2. Alexander, "Patton Film 'did not do him justice.'"

3. See Hirshson, *General Patton*, 635–36.

Chapter 24

1. It would be difficult to overestimate the importance of the role of Adolph A. Hoehling, The Army Times Publishing Co. book editor. Without his persistent efforts, *G–2: Intelligence for Patton* might never have been published.

2. *G–2: Intelligence for Patton* was reissued in a paperback ed. in 1999 by Schiffer Publishing Co. in its Military History series. As of this writing, it is still in print.

3. Lt. Col. Albert N. Garland, "Reviews and Recommended Reading," *Infantry*, 62 no. 3 (May–June, 1972), 61; Brig. Gen. Donn A. Starry, *Armor*, March–April, 1972, 66.

4. Col. James M. Hillard, "Right of Center," *Wilson Library Bulletin*, 47 no. 2 (Oct. 1972), 217. Col. Wilson was the librarian at The Citadel, the Military College of South Carolina, Charleston.

5. Essame, *Patton: A Study in Command,* 122.

6. Charles B. MacDonald, *A Time for Trumpets: The Untold Story of the Battle of the Bulge,* Quill ed. (New York: William Morrow, 1985), 69.

7. Geoffrey Perret,: The *There's A War to Be Won: The United States Army in World War II* (New York: Ballentine Books, 1997), 399.

8. Charles M. Province, *Patton's Third Army*, Hippocrene paperback ed. (New York: Hippocrene Books, 1994), 116.

9. Carlo D'Este, *Patton: A Genius for War* (New York: Harper Collins, 1995), 676.

10. George Forty, *The Armies of George S. Patton* (London: Arms and Armour Press, 1998), 218–19.

11. Hirshson, *General Patton*, 570.

12. William E. Odom, "Testimony before the U.S. Senate Select Committee on Intelligence," July 20, 2004, 3. Odom's Koch reference was from Harold C. Deutch, "Commanding Generals and Uses of Intelligence," Intelligence and National Security, no. 3 (July 1988), 194-261.

13. Ambrose, *Citizen Soldiers*, 180

14. *Ibid.*, 200–208.

15. John S. D. Eisenhower, "Patton and the Battle of the Bulge," *The Patton Saber* (Fort Knox, Ky.: The Patton Museum and Foundation), Winter 2005, 1.

16. Lt. Col. Michael D. Rosenbaum, "The Battle of the Bulge: Intelligence Lessons for the Army after Next," Strategy Research Project, U.S. Army War College, 1999.

17. *Ibid.*, 25.

18. *Ibid.*, 14.

19. *Ibid.*, 21–2.

Chapter 25

1. "Patton" screenwriters Francis Ford Coppola and Edmund H. North also based their script on Bradley's book, *A Soldier's Story*, and Ladislas Farago's *Patton: Ordeal and Triumph*.

2. Conrad C. Crane, "Beware of Boldness," Parameters, U.S. Army War College Quarterly, 30 no. 1 (Summer 2006), 88.

3. Alexander, "Patton Film 'did not do him justice.'"

4. Maj. Kevin Dougherty, "Oscar Koch, An unsung hero behind Patton's victories—Our MI Heritage," Military Intelligence Professional Bulletin, 28 no. 1 (April–June, 2002), 66.

5. *Pyle, Brave Men*, 465.

Index

A General's Life (Bradley), 199
A Soldier's Story (Bradley), 25
A Time for Trumpets (MacDonald), 287
"A Wing and a Prayer" (Eddie Cantor), 64
Adjutant General's Corps, 252
Afghanistan, 261
Afrika Korps, 81, 87
Agrigento, Sicily, 142, 144, 155
Air photo reconnaissance, 129–130
Air University, 292
Airborne units: 101st Airborne Division, rescued at Bastogne, 28; moved from Fort Jackson, 95; mentioned, 108, 273; 82nd Airborne Division, Sicilian invasion, 147–150
Alexander, Gen. Sir Harold R. L. G., 82
Alfred A. Knopf publishers, 52
Allen, John W., 17–18, 45, 75–77, 277
Allen, Robert S.: friendship with Koch, 50; wrote about Koch in *Lucky Forward*, 42, 50–51; cited by Koch as intelligence officer, 215–216; mentioned, *passim*
Alpine Redoubt, 227–229
Alumnus magazine, Southern Illinois University, 270, 271
Ambrose, Stephen, 203, 290
Anfa Conference, 127
Anzio invasion, 133
Apollo spacecraft, 269
Arlington National Cemetery, 280
Armies of George S. Patton, The (Forty), 289
Armor journal, 286
Armored units: I Armored Corps, 81, 128, 309; II Corps, 82, 154, 197, 309; 2nd Armored Division, 24, 26, 33, 41, 156; 6th Armored Division. 42, 70, 72, 259–260; 761st Tank Battalion, 106
Armstrong, Commander Neil, 269
Army Times, 196, 284–285
Arnold, Lt, Col. Henry ("Hap"), 40
Art of War, The (Sun Tzu), 261
Associated Press, 15, 234
Avaranches bridge, 176

"Bad Moon Rising"
(Credence Clearwater
Revival), 269
Baldwyn, Hanson W., 195
"Ballad of the Green Berets"
(Barry Sadler), 9
Bastogne, Belgium, 28–29,
206, 260
Battle of Khe Sanh, 224
Battle of the Bulge, 7, 28, 51,
60, 88, 94, 100, 125,
193–204, 209–219
Battle of Ypres, 263
Battles Lost and Won
(Baldwin), 195
Beatles, The, 269
Berry, Chuck, 98
Bigelow, Michael E., 220
Bitter Woods, The (John
Eisenhower), 272
"Black Book" on Sicily,
129–130
Blackburn, Col. Donald, 97
*Blackburn's
Headhunters* (Harkins),
97
Blackstone, Sub–taskforce,
2, 27, 79–80
Blair, Clay, 199
Blood on the Roses (Hays),
69
Bocelli, Andrea, 303
"Brady Bunch, The," 269
Brittany, France, 174, 183
Brown v. Board of
Education, 109
Browning, Robert, 144
"Butch Cassidy and the
Sundance Kid," 269

Caddick–Adams, Peter, 349
Caldwell County, Kentucky,
21
Caldwell, Isaac, 22
Caldwell, John, 21
Calley, Lt. William. 268
Caltanissetta, Sicily, 137
Camp Lejeune, North
Carolina, 241–243
Cantor, Eddie, 64
Capital Times, Madison,
Wisconsin, 9
Carbondale, Illinois:
location 13, 21; citizen's
advisory board, 31;
Memorial Day
Association, 14–17, 77;
Rotary Club, 305
Carmi, Illinois, 94
Carter, Lt. Col. Bernard S.,
33
Casablanca, Morocco,
79–80, 128
Catania Plain, Sicily, 138
Cavalry Journal, 40
Cavalry School, U.S. Army,
41, 88, 215, 231, 308
Calvary units: 1st Wisconsin
Cavalry, 38; 14th
Cavalry, 73; 15th
Cavalry Squadron, 186
Champaign, Illinois, 50
Chanute Field, Illinois, 112
Charles, Ray, 98
Chattanooga, Tennessee, 21

Cherbourg Peninsula, France, 176
"Cherry, Cherry" (Neil Diamond), 103
Chicago, Illinois, 21, 37, 108, 109, 225
Chicago Tribune Magazine, 211
Chicago White Sox, 12
Choate, Clyde, 15
Churchill, Prime Minister Winston S., 128, 195
CIA, Koch's association with, 10, 236–238, 248
Citadel, The, 167
Citizen Soldiers (Ambrose), 203, 290
Civilian Conservation Corps, 39, 40
Clark, George Rogers, 21
Clayton, Melinda, 350
Cobb, Ty, 12
Codman, Charles R., 55
Coffin, Rev. William Sloan, 224
Cole, Nat King, 249
Columbia College, South Carolina, 109
Columbia, South Carolina, 94–96, 98
Combat infantry training, 69–72
"Comfortably Numb" (Pink Floyd), 303
Comiso, Sicily, 154
Command and General Staff School, U.S. Army, 215

Command support, 244–245, 294, 306
Congaree Air National Guard Base, South Carolina, 242
Corley, Mary, 98, 134, 135, 250. See also Mary Hays
Cornell University Press, 52
Corsica, 136
Costello, Gen. Normando A., 72, 98–99
Costello, Toni, 99
Crane, Conrad C., 296
Creedence Clearwater Revival, 269
Croesus of Lydia, 118. See also Oracle of Delphi
Cross, Sgt. Jesse, 101
Crusade in Europe (Dwight Eisenhower), 194

D'Este, Carlo, 220, 288
D–Day landings, France, 173–176
"Departmental Ditties" (Kipling), 258
Des Moines, Iowa, 39
"Desert Fox." See Rommel, Field Marshal Erwin
Desert Training Center, 27
Detroit Tigers, 12
Deutsch, Harold C., 290
Devers, Gen. Jacob L., 231
Diamond, Neil, 112, 303
Dickson, Col. Benjamin A. ("Monk"), 82
Dilliard, Irving, 76

Dobkin, Lawrence, 295
Domino, Fats, 98
Doolittle, Col. James H., 65
Doubleday publishers, 54
Dougherty, Maj, Kevin, 299
Drive (Codman), 55
Dukws, amphibian, 133–135

East St. Louis, Illinois, 40
"Easy Rider" movie, 269
Egyptian, The, Southern Illinois University student newspaper, 252
Eisenhower, Gen. Dwight D.: failure to act on Koch's pre–Bulge intelligence, 8, 60, 192–197; president, 108, 250; shut off Patton's fuel supply, 183–189
Eisenhower, John S. D., 271, 291
El Guettar, Tunisia, battle of, 86, 146
Enigima machine, 163. See also: ULTRA
Esquire magazine, 212
Eureka, California, 262
Evansville, Indiana, 64

Falaise gap, 177
Farago, Ladislas, 55–60, 124, 177, 196, 299
Fifth Army, U.S., 68
Fats Domino, 98
Fink, John, 211
First Army, U.S., 173, 175, 183, 196, 291
First Christian Church, Carbondale, Illinois, 31
Fort Benning, Georgia, 24, 41
Fort Bliss, Texas, 39
Fort Bragg, North Carolina, 79
Fort Campbell, Kentucky, 95, 273
Fort Crook, Nebraska, 73
Fort Des Moines, Iowa, 73
Fort Drum, New York, 292
Fort Huachuca, Arizona, 294
Fort Jackson, South Carolina, 73, 93–100, 108–110, 134, 248–250
Fort Knox, Kentucky, 291
Fort Leavenworth, Kansas, 215
Fort Leonard Wood, Missouri, 69–72, 98, 249, 258–260
Fort McDowell, California, 24
Fort Riley, Kansas, 24, 39, 88, 215, 231
Fort Sam Houston, Texas, 173
Forty, George, 289
Freedom Riders, 112
French North African 4th Tabor of Goums, 153

G–2: Intelligence for Patton (Koch with Hays), planning for, 47–52;

POW stories in, 89; on Sicily, 138–139; on cross-channel invasion, 174–175; on the Battle of the Bulge, 198, 208; on qualities of a good intelligence officer, 216; on Hitler's Alpine Redoubt, 228; on command support, 245

Gay, Gen. Hobart, 144

Gehlen Organization, German, 235

Gela, Sicily, 137–139, 148, 150, 154

Gell–Mann, Murray, 270

Gelman, Ben, 270

General Motors, 243

General Patton: A Soldier's Life (Hirshon), 33

Gerber, Capt. Helmut, 217

German Cavalry in Belgium and France, 1914, The, 36

German military units: Fifth Panzer Army, 82; Sixth Panzer Army, 200, 289; 10th Panzer Division, 86; 15th Panzer Division, 154; 16th SS Panzer Division, 228

GI Bill, 251

Gilmour, David, 303

Godfather, The (Puzo), 269

Grand Army of the Republic, 15

Granite City Press–Record,

Granite City, Illinois, 265

"Great Pretender, The" (The Platters), 301

Greenville–Spartanburg, South Carolina, 95

Ground General School, U.S. Army, 25, 231

Grow, Gen. Robert W., 41, 259–261

Guggenheim Fellowship, 17, 248

Hackett, Bobby, 73, 249, 303

Haines, Gen. Peter C. III, 233

Hardscrabble School, 76

Harmon, Gen. Ernest N., 27

Harster, SS Lt. Gen. Wilhelm, 121

Hart, Capt. B. Liddell, 52, 195

Harvard University, 268

Hawaii Division, 25

Hays, Alan, 21

Hays, David, 350

Hays, Mary, 32, 48, 59, 120, 122, 251, 278, 281, 302–305. See also, Mary Corley

Helfers, Maj. Melvin C., 125, 164–168

Hendrix, Jimi, 113

Herb Alpert and the Tiajuana Brass, 18

Herman Goering Division, 153

Hill, Lee, 264

Himmler, Heinrich, 228

Hirshon, Stanley P., 32, 289

Hitler, Adolph, 66, 80, 86–88, 100, 176, 182, 189, 194, 195, 198, 201, 227–229
Hodges, Gen. Courtney, 183
Hoehling, Adolph A., 281
Holmes, Justice Oliver W., 280
Hose, Warrant Officer Fred, 200, 217
Hotel Pfister, Milwaukee, 38
Humphrey, Hubert, 225
Huntsville, Alabama, 21
Husky, Operation, 128–133

I Was With Patton (Lande), 200
Il Divo, 303
Iliad of Homer, 262
Illinois Central Railroad, 11, 275
Illinois National Guard, 275
Illinois Ozarks, 21
Illinois State Historical Society, 17
Illinois State Police, 275
Infantry units: 1st Infantry Division, 53; 25th Infantry Division, 22;; 81st Infantry Division, 258; 99th Infantry Division, 203
Infield, Glenn B., 238
Intelligence officers, qualities of, 216
Intelligence, military: in early U.S. Army, 24–25; differences in between U.S., British, 117, 133

Invasion planning: North Africa, 78–80; Sicily, 128–138; Operation Overlord, 173–176
Iraq, American involvement in, 261
Iron Cross medal, 90
Italian military units: 26th Assieta Division, 155; 207th Coastal Division, 155
Ivan Obolensky publishers, 52, 58–60

Jackson County, Illinois, Historical Society, 32, 77, 310
Jane Addams Book Shop, 50
Jefferson Barracks, Missouri, 40
Jodl, Gen. Alfred, 63
Johnny Got His Gun (Trumbo), 265
Johnson, President Lyndon B., 224
Joplin, Janis, 113

Kairouan, Tunisia, 147
Kasserine Pass, 82
Kennedy, President John F., 280
Kennedy, Sen. Robert, 224, 225
Kent State University, Ohio, antiwar protests, 275
Kern, Edward, 211
Keyes, Gen. Geoffrey, 232

King, Dr. Martin Luther Jr., 105, 224
Kipling, Rudyard, 258
Klein, Chaplain Ernest C., 15
Klein, Gen. John A., 247
Knoxville, Tennessee, 110
Knutsford, England, 171
Koch, Gen. Oscar: and "Patton" movie, 6–7, 296; association with CIA, 10, 235–237; battle with cancer, 8, 122–123, 161–162, 179, 207, 211, 245, 246; death, 276; early career, 22–27; eulogy to, 307–311; "Jewishness," 35; modesty, 47–48, 142; name, 34–35; physical description, 22; reaction to Patton's death, 230; patriotism of, 253, 273; relationship with Patton, 20, 26–27, 103, 115. 116, 141–144, 168–169, 218–221, 230; reluctance to place blame, 203; view of Patton slapping incident, 145
Koch, Nannie Caldwell, 11, 20–23, 35, 50, 123, 143, 161–162, 214, 245, 276
Kreuger, Gen. Walter, 215

Lande, D. A., 200
Lanza, Mario, 98
Lawson, Capt. Ted W., 65
Legends and Lore of Southern Illinois (Allen), 76
LeMay, Gen. Curtis E., 225–226, 240
Licata, Sicily, 136, 154
Life and Death of Lizzie Morris, The (Hays), 250
Life magazine, 211
Light Horse Squadron Association, 23, 35–38, 76, 274
Lipizzan Stallions, 281
Little Richard, 98
Little Rock, Arkansas, 21
Logan, Gen. John A., 14–15, 273
Look magazine, 211
Lopez, Michael, 264
Lucky Forward (Allen), 42, 50–51
Luman, Col. Ralph M., 33
Luxembourg, 200, 248
Lyons, Bill, 278

MacDonald, Charles B., 287
Macmillan publishing, 52
MacVicar, Robert W., 275
Maddox, Col. Halley G., 142
Maddox, Governor Lester, 121
Maginot Line, 187
Malmedy, Belgium, 199, 238
Manson, Charles, 270
March Field, California, 40
Marine Corps units: 1st Marine Division, 258

Market Garden, Operation, 163, 240
Marshall, Gen. George C., 137
Marshall, Justice Thurgood, 121
"Mash" television series, 96
Maxey, David, 211
McCaffrey, Gen. William J., 269
McCarthy, Frank, 5–7, 295
McCarthy, Sen. Eugene, 224
McDowell–Obolensky publishers, 52
McGraw–Hill publishers, 52
Medal of Honor, 15, 60, 133. 299
Medicare, beginning of, 113
Memorial Day, origin of, 14–16, 273
Memphis Naval Air Reserve Training Center, 273
Memphis, Tennessee, 224
Messina, Sicily, 138, 156–160
Metz, France, 89, 174, 181–187
Meuse River, 183, 194, 239
Michigan, University of, 40
Michner, James A., 275
"Midnight Cowboy" movie, 269
Military Chaplain, The, journal, 207
Military History Institute, 292
Military Intelligence Hall of Fame, 2, 294

Milwaukee Auditorium, 38
Milwaukee Braves, 113
Milwaukee, Wisconsin, 23, 35–38, 76, 108, 113, 274, 307
Mississippi River, 21
Missouri Ozarks, 95
Missouri, University of, 265
Moe, Henry Allen, 53–54
Monrovia, U.S.S., 146
Montgomery, Gen. Sir Bernard L., 156, 177, 183, 240, 289, 296
"Mood Indigo" (Bobby Hackett), 303
Morocco, 28, 34, 79, 81, 128, 130, 309
Morris Library, Southern Illinois University, 77
Mt. Adams, Washington, 40
Muller, Col. Walter J., 186
Mussolini, Benito, 80, 86, 87, 157, 237
My Lai massacre, Vietnam, 268

Nancy, France, 193
NASA, 269
Nashville, Tennessee, 110
National Security Agency, 289
Native Americans, slaughter of, 262
Nat King Cole, 249
New York City, 24, 106
New York Times, 195, 262
Nixon, President Richard, 268

North Africa, invasion of, 2, 26–27, 78–81, 130

O'Brien, Bill, 278
O'Neill, Col. James H., 206–208
Obolensky, Ivan, 57, 59
Odom, William E., 289–290
Odenweller, Ted, 350
Ohio River, 21
Okinawa, 253
Old Main, Southern Illinois University, arson of, 267. 270–271
Oracle of Delphi, 118. See also, Croesus of Lydia
Orangeburg, South Carolina, 224
OSS, 131, 154, 158, 201, 239
Overlord, Operation, 173
Oxford, Mississippi, 21
Ozark Mountain range, 72

Pack animals in Sicily, 234
Paducah, Kentucky, 21
Palermo, Sicily, 139, 154, 156–160
Panama Canal, 24
Pancho Villa, 34, 38, 308
Paris, liberation of, 178, 182
"Patton" movie, 6–7, 295–297
Patton: A Genius for War (D'Este). 220, 288
Patton: A Study in Command (Essame), 286–287

Patton, Beatrice, 50, 297
Patton: Ordeal and Triumph (Farago), 60, 124
Patton's Third Army (Province), 288
Patton, Col. George S. III, 272, 281
Patton, Gen. George S. Jr., warns of attack on Pearl Harbor, 26; origin of nickname "Old Blood and Guts," 27; view of POWs, 88; and racism, 106; praise of Oscar Koch, 141, 142; slapping incident, 145; letters to soldiers of U. S. Seventh Army, 146, 160; battle tactics, 163–164, 218–220; on religion and prayer, 207–208; response to mistakes, 217–218; death, 230; sympathy for wounded soldiers, 261
Patton Hall, Fort Riley, Kansas, 231
Patton Museum Foundation, 291
Patton Prayer, 206–209
Patton Saber, 20, 69, 218
Patton Saber, The, newsletter, 291
Pearl Harbor, 26
Pearson, Drew, 50
Pelieu, battle for, 258
Peover Hall, Knutsford, England, 171
Perret, Goeffrey, 288

Pershing, Gen. John J., 34
Pink Floyd, 303
Point Olivo Airport, Sicily, 148
Polanski, Roman, 270
Porter, Gen. Robert W. Jr., 53
Praeger, Frederick A., 52
Prentice-Hall publishers, 54
Prisoners of war, 88–90, 150, 154, 159, 226
Propaganda. See psychological warfare
Province, Charles M., 288
Psychological warfare, 86–87, 150–152
"Purple Haze" (Jimi Hendrix), 113
Puzo, Mario, 269
Pyle, Ernie, 133, 264, 303

Racial strife in 1960s U. S., 8, 105–113
Radio Free Europe, 56
Random House publishers, 56
Rantoul, Illinois, 112
Ratisbon, Germany, 143. See also, Regensberg
Ravanausa, Sicily, 136
Rbia River, 78–81
Reagan, President Ronald, 113, 289
Red Badge of Courage, The (Crane), 263
Red Ball Express, 183
"Red Tails" movie, 112. See also, Tuskegee Airmen
Regensberg, Germany, 148
Reims, France, 63, 182
Republic Aviation Company, 64
Rhine River, 143, 183, 188
Rhodes, Governor James, 105
Richards, Frank, 263
Richmond, Virginia, 21
Rolling Stones, 123
Rolling Together newsletter, 209
Rommel, Field Marshal Erwin, 81–86
Roosevelt, President Franklin D., 127
Rosenbaum, Lt. Col. Michael, 292–294
Rumor as propaganda, 152

Sadler, Sgt. Barry, 9. See also, "Ballad of the Green Berets"
Safi, French Morocco, 79, 130
Saumur, France, artillery school, 39
San Antonio Express newspaper, 194
San Fratello, Sicily, 158
Sardinia, 136
Schiffer Publishing Company, 292
Schimpf, Gen. Richard, 188
Scott Air Force Base, Illinois, 273

Scott, George C., 295
Scott, Paul, 268
Second World War, The (Churchill), 195
Senate Select Committee on Intelligence, U.S., 289
Service stars, 63
"Sesame Street" television series, 269
Shawnee National Forest, 14
Shockey, Lt. Cyrus R., 42
Shryock Auditorium, Southern Illinois University, 224
Shwedo, Maj. Bradford J., 163, 188
Sibert, Gen., Edwin L., 197, 200, 203
Sicilia Liberata, American propaganda newspaper, 152
Siegfried Line, 183, 187–189
Skorzeny, Col. Otto, 100, 237–240
Slaughterhouse Five (Vonnegut), 269
Sledge, E. B., 253, 258
SmallWorld magazine, 76
Smith, Mrs. Burrell, 11–12
Smith, James B., 12
Smith, Gen. Walter B. ("Beetle"), 65, 236
Snipes, Macy Yost, 107
South Carolina Air National Guard, 241
South Carolina, University of, 94, 98, 109

Southern California, University of, 33
Southern Illinois University, antiwar protests at, 270, 274–276; mentioned, 10, 16, 61, 77 100, 214, 223, 224, 243, 251, 278
Southern Illinoisan, The, newspaper, 270
Spock, Dr. Benjamin, 223
Sprigle, Ray, 107
SS (*Schuzstaffel*), 228, 238
St. Louis, Missouri, 94, 265
St. Louis Post–Dispatch, The, 9, 18, 22, 28, 29, 45
St. Vith, Belgium, 199
Stackpole publishers, 54
State Department, U.S., 10, 236, 248
Sternberg, Vernon, 61
Strait of Messina, 157
Strong, Gen. Kenneth, 60, 200
Sun Tzu, Gen., 261
"Surrender, Hell," movie, 97
Susan B. Anthony Hall, Southern Illinois University, 214, 268
Sousloparov, Gen. Ivan, 63
Sykes, Sgt. Mason, 96, 242, 248

Tate, Sharon murder, 270
"Teller of Tales" magazine article, 76. See also: John W. Allen
Tenth Fleet, The (Farago), 55

The Platters, 301
There's a War to be Won (Perret), 288
Third U.S. Army: history, 42; war room, 101; and OSS, 131; pre–invasion planning, 171–175; stalled for lack of fuel, 183–189; in the Battle of the Bulge, 192–203
Thirty Seconds over Tokyo (Lawson), 65
Thunderbolt (P–47) fighter plane, 64, 178, 300
Till, Emmett, murder, 109
Torch, Operation, 2, 27
Traveler's Rest, South Carolina, 111
Troina, Sicily, 157, 158–160
True magazine, 209
Truman, President Harry S., 107, 113
Trumbo, Dalton, 265
Turner, Tim, 267
Tuskegee Airmen, 112

ULTRA, 10, 124, 163–169, 177, 183, 201
Ultra Secret, The (Winterbotham), 124
United Press International (UPI), 15, 121
U.S. Army Air Corps, 40
U. S. Marines, 75, 108, 241, 253, 278
USS Monrovia, 146

Vanguard Press, 52

Verdun, France, 199
Veterans Hospital, Marion, Illinois, 276
Von Palus, Field Marshal Friedrich, 86
Von Poseck, Gen. Maxmilian, 36
Von Runstedt, Field Marshal Karl Gerd, 188, 194
Vonnegut, Kurt Jr., 269

Wallace, Alabama Gov. George, 235
Wallace, Maj. Warrick, 164, 168
Walnut Grove School, 64, 300
Walter Reed Army Hospital, 161
War as I Knew It (Patton), 51
War College, U.S. Army, 53, 92
Warrior: the Story of Gen. George S. Patton, Jr. (Army Times editors) 196, 284
"Washington Merry–Go–Round," syndicated newspaper column, 50
Washington, D.C., 55, 121, 123, 161, 236, 248, 284
Watertown, South Dakota, 39
Wayne, John, 297
Wedemeyer, Gen. Albert, 138–139
Western Task Force, 26, 81

West Point, 38, 54, 99, 203
White County, Illinois, 67
"White Rabbit, The." See
    Yeo–Thomas, Forrest
Whitmore Publishing
    Company, 285
Wilson, President Woodrow,
    38
Winterbotham. Group Capt.
    Fredrick W., 124, 201
With the Old Breed
    (Sledge), 253
Wolf, Dr. Richard, 302
Woodlawn Cemetery,
    Carbondale, Illinois, 12,
    14–16, 21, 273–274

XV Corps, 177

Yeo–Thomas, Wing
    Commander Forrest,
    240
Young, Walter Sr., 67, 252

Zimmerman, Dr. Murray, 33,
    100, 116, 123, 143–144,
    237–238, 245
Zimmerman, William, 37

About the Author

Robert Hays is the author of seven novels and a book of short stories and has written, edited, or collaborated on a half–dozen works of non–fiction. His short stories have appeared in anthologies and he has published numerous academic journal and popular periodical articles. Selections from three of his novels gained Pushcart Prize nominations. He is a U.S. Army veteran and, though retired from classroom teaching, holds professor emeritus rank at the University of Illinois. He has received international awards for both his teaching and research and was selected twice as a C–SPAN visiting professor. He also taught in Texas and Missouri and was on the faculty of Southern Illinois University. He lives in the beautiful southern Illinois wooded hill country about which he often writes.

About Brock Ayers

Major Brock Ayers, U.S. Army (ret.), is a former intelligence officer who served on the National Security Council / White House Situation Room staff under presidents Ronald Reagan and George H. W. Bush as well as a two-year stint in the Defense Intelligence Agency's space and special operations section. On various tactical assignments he was a unit intelligence officer (S2) and liaison officer to the Republic of Korea Army, Japanese Ground Self Defense Force, and the Australian Army. A vice president of a national investment firm, he also is a National Churchill Museum Fellow and was a 21-year member of the Westminster College board of trustees.

Author's Note and Acknowledgments

I have made no attempt to update the basic factual information from the original edition of this work. It is a study set in the past, and facts and interpretations relative to General Oscar Koch have not changed. To the extent there is anything new, it simply is that General Koch and his accomplishments have come to be more widely known and more fully understood and appreciated. Time cannot erode his stature. Today, or even years into the future, his story will be the same.

From the outset, I have drawn on the work of some of the premier military historians and intelligence professionals to support my own highly subjective analysis of General Oscar Koch's performance as an intelligence officer. They are named in the text so readers need not go to obscure endnotes to learn who they are. I am very much indebted to them individually and collectively.

I could add a great many such references. For example, the eminent British scholar, Peter Caddick–Adams, whose 2015 Oxford University Press book, Snow & Steel, is seen by many as the definitive study on the Battle of the Bulge, drew heavily on General Koch's work and proclaimed him to deserve a "greater share of the limelight."

He wrote me, "I found that researching George S. Patton through the eyes and actions of General Oscar Koch was the most satisfactory approach to understanding Third Army in the Bulge," To this end, he used both G–2: Intelligence for Patton and the present work as sources.

I thank Schiffer Publishing Company for its benevolent permission to quote extensively from G–2: Intelligence for

Patton, to which it reserves all rights. Schiffer has kept this book in print for a number of years and, so far as I know, plans to do so for some time to come.

Major Brock Ayers, U.S. Army (ret.), who no doubt knows more about military intelligence than I ever could learn, has been a gracious advocate for this book and kindly consented to write a perceptive foreword to this new edition. My very great appreciation to him.

Members of my Champaign writers group offered important support and encouragement. I am grateful to them all, but especially to Ted Odenweller, a U.S. Air Force veteran and "military brat" whose father was a career officer, who fully understands military structure and the critical nature of intelligence and was particularly helpful.

I thank my son, David Hays, for his careful reading of parts of the text and helpful suggestions. He brought to the table the skilled analytical and objective consideration of facts developed in his own years as a journalist.

Melinda Clayton at Thomas–Jacob Publishing LLC superbly guided this new edition to completion. I could never say enough about her dedication and her usual skilled work, nor adequately express my gratitude.

My personal circumstances are quite different than they were at the time the first edition of this work was published. I have retired from active teaching and moved from our long–time home in Champaign, Illinois, to my home area of southern Illinois. Sadly, the move was precipitated by the illness of my wife, Mary Corley Hays. She suffered Alzheimer's disease for eight years and passed away in March, 2021. But, like the story of Oscar Koch, all mentions of her in this book are and always will be authentic.

She stood by patiently through all the disruptions of

normal life brought on by this and all my writing. She personally had great admiration and respect for General Oscar Koch and was among the most enthusiastic supporters of my goal to tell the world who he was. Without her, this book never would have been written.

Other Books by Robert Hays

Non–Fiction

Editorializing "the Indian Problem"
A Race at Bay
State Science in Illinois
G–2: Intelligence for Patton (with Gen. Oscar Koch)
Country Editor

Fiction

An Empty House by the River
An Inchworm Takes Wing
A Shallow River of Mercy
Blood on the Roses
The Baby River Angel
The Life and Death of Lizzie Morris
Circles in the Water
Equinox and Other Stories
Early Stories from the Land (editor)

www.ingramcontent.com/pod-product-compliance
Lightning Source LLC
Chambersburg PA
CBHW070507240426
43673CB00024B/472/J